Wetlands Ex

Wetlands Explained

Wetland Science, Policy, and
Politics in America

WILLIAM M. LEWIS, JR.

OXFORD
UNIVERSITY PRESS

2001

OXFORD

UNIVERSITY PRESS

Oxford New York
Athens Auckland Bangkok Bogotá Buenos Aires Cape Town
Chennai Dar es Salaam Delhi Florence Hong Kong Istanbul Karachi
Kolkata Kuala Lumpur Madrid Melbourne Mexico City Mumbai Nairobi
Paris São Paulo Shanghai Singapore Taipei Tokyo Toronto Warsaw

and associated companies in
Berlin Ibadan

Copyright © 2001 by Oxford University Press, Inc.

Published by Oxford University Press, Inc.

198 Madison Avenue, New York, New York 10016

Oxford is a registered trademark of Oxford University Press.

Library of Congress Cataloging-in-Publication Data
Lewis, William M., 1945–
Wetlands explained : wetland science, policy, and
politics in America / William M. Lewis, Jr.
p. cm.
Includes bibliographical references.
ISBN 0-19-513183-5; ISBN 0-19-513184-3 (pbk.)
1. Wetlands. 2. Wetland conservation—
Government policy—United States. I. Title.
QH87.3 L49 2001
578.768—dc21   2001039091

1 3 5 7 9 8 6 4 2

Printed in the United States of America
on acid-free paper

PREFACE

Wetland science is a new and rapidly evolving branch of ecosystem science, and wetland regulation is a new and rapidly evolving sociopolitical enterprise. The two have an intense relationship that is in many ways reminiscent of the relationship between nuclear physics and national defense 50 years ago. Regulatory initiatives constantly raise unanswered scientific questions, while scientific study supports or calls into question regulatory practice; both regulation and science develop in an atmosphere that is highly charged politically.

With a few notable exceptions, writings on wetlands have been directed to individuals who have special knowledge of some aspect of wetland science, policy, or regulation. Concise overviews of the entire field without the presumption of special knowledge are difficult to find, and yet they are essential in broadening the general accessibility of this inherently multidisciplinary subject. The purpose of this book is to bring together, in compact form, a broad scientific and sociopolitical view of U.S. wetlands, without assuming that the reader has a specialized background.

This work stems from my association with the National Research Council's Committee on Wetland Characterization (1994–1995). The committee's report, which is cited at numerous points in this book, is my point of departure in trying to make this diverse subject matter as transparent as possible to nonspecialists. I am grateful to the members of the NRC committee; many of them have given me valuable information as I worked on this book. I am especially indebted to former committee members M. Strand, F. Bosselman, and C. Johnston, and I thank David Cooper for the cover photograph. Among others who have helped me are P. Diggle, J. Gosselink, F. Dahm, M. Davis, S. David, R. Bernstein, J. Kitchell, and J. Kusler, all of whom I exonerate from errors that I may have made in using what they have given me.

Boulder, Colorado   W. M. L. Jr.
1 August 2000

CONTENTS

Wetlands Explained

1

WHERE WE ARE, AND
HOW WE GOT HERE

English is a subtle language with many words that offer fine shades of meaning, but it also can be blunt and unequivocal. Dictionaries were not made for words such as hairdo, ballpark, or pigpen. The law, however, as practiced by Americans, can mutate the meaning of even the humblest word. If the law concerns itself with pigpens, then we must know whether a pigpen still exists when the pigs are removed and, if so, for how long. We must know if a pen originally built for cattle can become a pigpen if occupied by pigs and if pigpens are the same in all parts of the nation. In short, we must have federal guidance, regional interpretations, legal specialists, and technical authorities on pigpens. So it is with wetlands.

The chapters of this book will show how troublesome the definition of wetlands has become since the federal government began regulating them. In the meantime, it will suffice to define wetlands informally as those portions of a landscape that are not permanently inundated under deep water, but are still too wet most years to be used for the cultivation of upland crops such as corn or soybeans. Wetlands, in other words, coincide pretty well with the common conception of swamps, marshes, and bogs.

The Eyes of the Beholder

Government has had its hand in wetlands for about 150 years. Between the 1850s and 1970s, the federal government was intent on eliminating

3

wetlands. Since then, it has been equally intent on preserving them. An individual who behaved in this manner would seem at least irresponsible. Many critics of federal wetland policy have in fact given the government a sound thrashing for its inconsistency, but the shift from elimination to protection of wetlands has continued nevertheless.

Blaming government is the duty of a free people, and also good sport. Even so, the obvious truth about wetland regulation is that government has merely reflected a change in public attitude toward wetlands. Most Americans now believe that wetlands should be saved throughout the nation, except possibly on their own property. Americans did not always feel this way.

Most European colonists of North America came from homelands that were essentially tame. By the middle of the eighteenth century, much of the European landscape was either plowed or grazed, and eradication of forests had been in progress since Neolithic times.[1] English wetlands were progressively diked in the Middle Ages to make way for grazing and cultivation and were drained on a massive scale beginning in the last half of the seventeenth century.[2] In contrast, the North American landscape that Europeans colonized was as wild as any on Earth. What is now the conterminous United States originally included 220-million acres (now approximately 100 million) of swamp, marsh, and bog, even before the subsequent addition of Alaska's 170-million acres of wetlands (Dahl 1990). Swamps extended broadly along many rivers, and the uplands were speckled with small swamps and marshy pockets. The United States also had several wetlands of global significance, including the Everglades, which persist today in altered form; the bottomlands of the lower Mississippi, which are now reduced but still extensive; and the Kankakee marsh, which once covered a large portion of upper Illinois and Indiana but has now essentially disappeared (figure 1-1).

1. According to observer Tobias Smollett, the British landscape was, by the 1760s, "smiling with cultivation . . . parceled out into beautiful enclosures" (Briggs 1983).

2. Bosselman's (1996) research showed that the diking of common wetlands to prevent the entry of floodwaters occurred even prior to historical record and that a profitable escalation of wetland conversion through drainage in the second half of the seventeenth century financed by the English aristocracy led to progressive restriction of public access to wetlands. Bosselman speculates that the drainage effort, which eventually was accelerated with the introduction of the steam engine, may have been inspired by the great success of the Dutch in draining and diking inundated lands on a large scale.

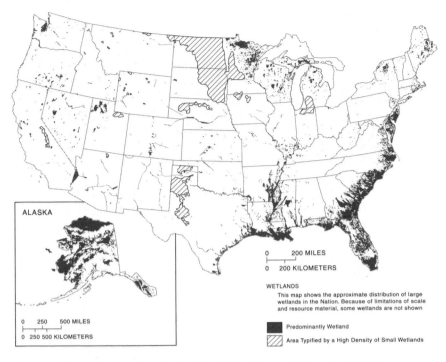

Figure 1-1. Distribution of wetlands in the United States (Winter et al. 1998).

Wetlands were a nuisance to colonists, as they had been in Europe.[3] While they produced fish, fur, fowl, and fiber, they also impeded transportation and were impossible to cultivate, except in the rice-growing regions. When drained, however, many wetlands yielded rich soil capable of sustaining high yields of crops that tapped centuries of natural nutrient accumulation. Thus, wetlands were viewed as undeveloped agricultural resources, as they had been in Europe.

In 1850, the U.S. Congress passed the Swamp Land Act, which

3. Wetlands were designated during medieval times and well afterward as "waste" (Bosselman 1996). The modern connotation of this term may be misleading, however, in that the wholesale conversion of wetlands in England generated massive political unrest as the benefits of wetlands were lost to large numbers of people. The growing public claim of control over private wetlands in the United States is in a sense a reversal of the sequence of events that occurred in England during the seventeenth century.

was intended to encourage the conversion of wetlands to agriculture or other uses. Congress reasoned that it could encourage the conversion of "swamp and overflowed lands" to agricultural use by ceding them to the states, which would through their own initiative or that of their citizens be able to finance a conversion to agricultural use. A total of 64-million acres eventually passed to the states in this manner. The identification of these lands was so loose as to make the present procedures for identification of wetlands seem an exact science (Bosselman 1996).

The Swamp Land Act of 1850 and its reformed successor (1855) were ineffective in making the large-scale conversions that were originally intended by Congress. The swamp land acts did, however, foreshadow the federal and public mind-set toward wetlands, which was reflected over the next century or more in many schemes to subsidize and encourage the conversion of wetlands. Governmental and private efforts had reduced the total acreage of wetlands in the United States by about 50% as of the mid-1980s (Dahl 1990).

Under the Influence of Ducks

One of the earliest sustained efforts to protect wetlands was that of duck hunters, who have been numerous in North America since about 1870, when effective shotguns first became widely available. In fact, duck hunting for decades ranked with golf and five-card stud as a pastime among CEOs and lawmakers. Thus, when decline in migratory waterfowl populations became noticeable, some weighty political and financial forces began to support the preservation of wetlands. Biologists had made a connection between the decline of flyway populations and drainage projects (figure 1-2); one remedy for fading waterfowl populations seemed to lie in aggressive purchase and lease of habitat.

Federal efforts to sustain waterfowl populations center around the National Wildlife Refuge System, which evolved in cumulative fashion beginning near the turn of the century at the initiative of Presidents Harrison and Roosevelt. This system, which is actually a loose confederation of sites administered for various purposes by the U.S. Fish and Wildlife Service, comprises 442 units totaling 91-million acres, of which 76 million are located in Alaska. While managed according to various objectives, maintenance of migratory waterfowl populations

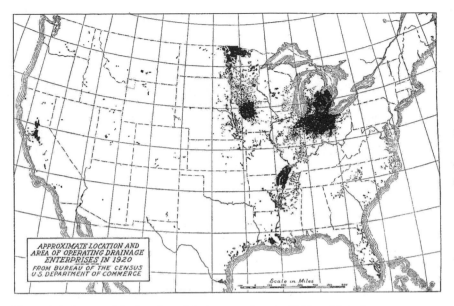

Figure 1-2. A map of the location of drainage projects in 1920, published as part of an effort to document the connection between decline of waterfowl and elimination of wetlands (Phillips and Lincoln 1930).

through habitat preservation is a major underlying motivation for the entire system. Legislation that has given longevity and direction to the system includes the Migratory Bird Treaty Act of 1918 and the Migratory Bird Conservation Act of 1929, the latter of which explicitly recognized the need for acquisition and preservation of habitat. Financial support for the system in recent times has come from the Migratory Bird Habitat Stamp Act of 1934 (duck stamp fee for hunters) and the Land Water Conservation Act of 1964, which authorizes the collection of user taxes and energy taxes for land acquisition (Fink 1994).

The continuing strength of interest related to duck hunting is well illustrated by the wetland conservation of Ducks Unlimited (DU), a nonprofit organization primarily oriented around waterfowl hunting. DU, which currently has a membership near 600,000, has raised approximately 1-billion dollars, much of which has been used to protect approximately 1-million acres of waterfowl habitat (Ducks Unlimited 1997). Conservation of wetlands also has been promoted by other kinds of organizations, such as the Izaak Walton League, whose members have found intrinsic value in wetlands and their biota, or have foreseen the value of wetlands to fisheries and recreation (Anfinson 1995).

Although duck hunters and conservationists, acting both privately and through the federal government, have protected numerous important wetlands, no system based on purchase or lease could possibly have protected more than a fraction of the 220-million acres of wetland originally found in the contiguous 48 states. Consequently, the progressive elimination of wetlands continued even after wetland advocates became well organized.

Congress Legalizes the Water Cycle

A steep increase in U.S. commitment to water quality occurred with passage of the Federal Water Pollution Control Act amendments of 1972, which have been known since their modification in 1977 as the Clean Water Act. This legislation, which sets as a goal the wholesale protection and restoration of chemical, physical, and biological "integrity" for the nation's waters, may be the most ambitious commitment ever made to a natural resource. The United States Environmental Protection Agency (EPA) has estimated that annual expense of the Clean Water Act exceeded 50-billion dollars as of 1990, and could be expected to increase steeply thereafter (EPA 1990). This expense is shared among federal, state, and local governments as well as industrial dischargers (it would be tiresome to mention taxpayers), and has been to a large extent accounted for by installation of wastewater treatment facilities. Although the cost of the Clean Water Act has a sobering effect even on water-quality enthusiasts, waters of the United States have a more nearly natural character chemically, aesthetically, and in capability for support of aquatic life as a result of the act (Patrick et al. 1992).

The present federal regulatory system for wetlands grew primarily out of the Clean Water Act, although the amendments of 1972 did not mention wetlands. The first priority in implementation of the Clean Water Act was to put a treatment plant behind every wastewater discharge pipe (*point source*) in the United States, which has been no offhand project. Complete protection of water, however, must also include dispersed influences (*nonpoint sources*) on water quality. Critical to the control of nonpoint sources is preservation of the natural cleansing function of wetlands.

In debating the Clean Water Act, Congress recognized that water overrides jurisdictional boundaries and missions of federal agencies. The key principle was enunciated by Senator Muskie and others, who declared during debate leading to the 1972 FWPCA amendments that water "moves in hydrologic cycles" (Bosselman 1996). This physical axiom, which subsequently appeared as a sort of mantra surrounding congressional debate, led to the inevitable implication of the Clean Water Act: The protection of water must extend to the margins of the drainage net, and thus to wetlands, even though wetlands are not mentioned specifically in the amendments of 1972 and appear only in a single phrase in the 1977 amendments of the Clean Water Act. The judicial system, interpreting congressional intent, reached a similar conclusion by its own means.[4]

Navigation in Shallow Water

Congress used the concept of navigable waters in setting the jurisdictional boundaries for the Clean Water Act and other federal legislation relevant to protection of surface waters.[5] The root of federal jurisdiction over navigable waters is federal authority over matters that involve multiple states or foreign nations, as navigation does. Aquatic navigation in the dictionary sense (i.e., travel by ship or boat), however, is not possible at the fringes of drainage basins

4. During 1975, in *National Resources Defense Council v. Callaway*, the Federal District Court of the District of Columbia ruled that the Army Corps of Engineers had not interpreted the breadth of its responsibilities in a manner consistent with the Federal Water Pollution Control Act amendments of 1972; the result was extension of regulatory authority over essentially the entire drainage net, with the possible exception of "isolated" waters. In the decision of *U.S. v. Riverside Bayview Homes* (1985), the Supreme Court upheld the authority of the Army Corps to exercise control over wetlands adjacent to (i.e., connected to) navigable waters for the purpose of protecting water quality (Strand 1993). In 2001, the U.S. Supreme Court seemed to take one step back from its earlier interpretation by denying protection for isolated wetlands when support of migratory birds is the main justification for protection (*Solid Waste Agency of Northern Cook County v. United States Army Corps of Engineers* [2001]).

5. The Army Corps of Engineers has federal authority under the Rivers and Harbors Act of 1890 and the Refuse Act of 1899, as since amended, to control the discharge of refuse into navigable waters. This authority was first used broadly for the protection of wetlands through President Nixon's Executive Order 11574 of December 23, 1970.

where surface waters first accumulate. Thus, proponents of the Clean Water Act had the option of either assaulting the principle of states' rights or expanding the legal extent of navigability; the latter seemed more feasible. In fact, it is sometimes said in the West, where the broadened concept of navigability reaches the outer limits of imagination, that navigable waters are those capable of floating a matchstick.

Navigable waters, which are now often designated by the more plausible synonym "waters of the United States," encompass for wetland regulation (but not necessarily for all other purposes) all surface waters having connections to waters that could be used in interstate and foreign commerce.[6] In the pockets and fringes of the drainage net and at the margins of flowing and standing waters, hydrologic connections typically involve wetlands. The Clean Water Act could not achieve its purpose without such an expansive interpretation of navigability because big waters collect from small waters. The nature of the drainage network, rather than legal principles, forces the jurisdiction of the Clean Water Act to encompass wetlands, as realized by Congress and by the courts.

Plugging a Leak

Section 404 of the Clean Water Act deals with wetlands but seems very selective in doing so. A sensible and straightforward way to protect wetlands would be to require that they be neither filled nor drained, given that water is the one thing that cannot be missing in a wetland. Section 404, which probably was written by people who

---

6. Authority comes from the so-called *Commerce Clause* of the U.S. Constitution: Article I, Section 8, Clause 3. Navigable waters are defined in the Clean Water Act as waters of the United States (Section 502[7]). Regulations of the Army Corps of Engineers define "waters of the United States" as waters currently or previously used for interstate or foreign commerce, transboundary waters including wetlands, interstate waters the degradation of which could affect interstate or foreign commerce, and tributaries or wetlands adjacent to or connected to waters of the United States as defined by interstate or foreign commerce (33 CFR 328.3[a]). Regarding matters of water quality, the expansion of regulations to the fringe of the drainage net is accomplished mostly by reference to the connectivity of waters within the net. The broadened definition of navigability for purposes of the Clean Water Act does not necessarily apply to other legislation having to do with navigability.

needed to be more concerned about legal and political probity than about sensibility, refers to the discharge of "dredged and fill material," and not to drainage.[7] Thus, a literal interpretation of Section 404 would suggest that a wetland can be drained as long as it receives no discharge of solids.

Here we enter the quagmire of intent and interpretation. If wetlands preserve water quality (which they do), and if wetlands are destroyed by drainage (which they are), then it makes no sense to ignore the issue of drainage in protecting wetlands. For this reason, Section 404 has been broadly interpreted to prohibit major tampering with any of the physical characteristics of wetlands, except when a permit has been issued for this purpose. Drainage seldom can be accomplished without ditching. Ditching involves removal of soil, which usually must be deposited beside the ditch.[8] Any good attorney can show a sympathetic judge that this act, even if involving only a shovelful of dirt, constitutes the discharge of fill material and thus comes under the prohibition of Section 404. In this way, the prohibition against filling has worked as a prohibition against drainage as well. Even so, wetlands sometimes can be effectively and legally dried up by diversion of water at a point remote from the wetland.

7. The Clean Water Act Section 301(a) seems to extend the intuitive definition of dredged and fill material by prohibiting the discharge of "any pollutant," with certain exceptions.

8. A critical legal landmark in the interpretation of drainage operations as related to prohibition of discharge occurred in a 1990 civil suit that pitted the North Carolina Wildlife Federation and the National Wildlife Federation against the Army Corps of Engineers (Army Corps) and the EPA. The plaintiffs claimed that the Army Corps and the EPA had allowed massive drainage of wetlands in eastern North Carolina without a permit by using a technical ruse that involved installation of drainage around wetlands, which led to the elimination of the hydrologic qualifications for their remaining classified as wetlands (*North Carolina Wildlife Federation et al. v. Tulloch* [1991]). The plaintiffs prevailed; the result was adoption of the Tulloch Rule by the Corps and EPA in 1993 (*Federal Register* 45093). The rule essentially broadened the definition of discharge to include any release of material associated with drainage or excavation. Even activities on a small scale must meet the test of avoiding alteration of a wetland such that it would no longer be classifiable as waters of the United States and that of avoiding degradation of wetland function. In 1998 the U.S. Court of Appeals for the District of Columbia ruled that the Clean Water Act could not be used in this way to prevent drainage of wetlands, but, more recently (January 2001), the EPA and Army Corps seem to have neutralized the court's decision by redefining some critical terms in the Tulloch Rule (http://www.epa.gov/owow).

The Fox and the Henhouse

Section 404 does not prohibit the alteration or elimination of wetlands. It merely requires that alteration or elimination be conducted under permit (404 permit). When Congress created the permitting requirement, it debated the assignment of responsibility for permitting. Some legislators held that the issuance of permits should be a responsibility of the Army Corps of Engineers. Their reasoning was that the Army Corps holds the lead responsibility for dredging and filling, and therefore should be in a good position to issue permits for these activities. This makes a certain amount of sense, although one could use a similar argument to prove that maternity wards should be run by morticians.

Others noted that the purpose of Army Corps activities related to dredge-and-fill is to promote navigation, which is not the purpose of Section 404. Underlying the formal arguments, of course, was a general feeling that the Army Corps would be less zealous (or less unreasonable, depending on one's point of view) than the EPA in applying Section 404. Ultimately, it was agreed that permits would be issued by the Army Corps and that the EPA would have the right of review.[9] Other agencies, especially the U.S. Fish and Wildlife Service (USFWS), would have the right to comment.

Environmentalists were incensed that the Army Corps, in their view the greatest despoiler of inland waters in the history of the Earth, would be responsible for the protection of wetlands. The henhouse analogy got an extraordinary workout while the Army Corps scrambled into position on both sides of the line of scrimmage.

Although the assignment of permitting to the Army Corps must have seemed ridiculous to many, hindsight has shown that the proponents of permitting by the Army Corps were either wiser or luckier than they appeared to be in the 1970s. If the chiefs of agencies have

9. The role of the EPA extends beyond mere review in the sense that the EPA interprets the standards under which permits are issued; the EPA decides whether the policies developed by the Army Corps for issuance of permits are consistent with the water quality goals that are described in the Clean Water Act. The standards used by the EPA in interpreting consistency between the permit program and the Clean Water Act include the concept of practicable alternatives (the permittee must show that there is no practicable alternative to conversion of wetlands) and the concept of significant degradation (degradation must be minor or must be offset through some replacement or restoration scheme [mitigation]).

anything in common, it must be that they know the value of attitude adjustment in surviving scrutiny by Congress and the public. The Corps adjusted (Shallat 1994).

The Army Corps of Engineers has proven itself competent and professional in 404 permitting.[10] This is not to say that the Army Corps does not make mistakes, but rather that the permitting program seems about as well off having been vested with the Army Corps as it would have been with another agency.

404: A Bone in the Throat

Permitting is a well-worn concept for Americans and American businesses. Most permitting schemes have been resisted strenuously at one time or another because they restrict freedom of action and usually cost money, either directly or through loss of efficiency. Many permitting schemes, such as those for building a house or developing land, are more pervasive and represent greater expenditure by far than the federal protection of wetlands. One may then ask why the 404 permitting scheme is viewed by many as the Mother of All Regulations. The answer lies in the paradoxical view of society that wetlands are both land and water.

Federal environmental regulations apply primarily to air and to water, and not to land. Because air and water do not remain fixed, it seems obvious that the common interest of society in air and water overrides private interest when the two are in conflict. In contrast, land stays put, and the balance of rights and responsibilities seems to run in the other direction; the owner of land should be able to use it with the broadest possible latitude, which would include alteration of its form, drainage, vegetative cover, and other attributes. This principle is still intact with respect to uplands (lands that are not wetlands).

Wetlands do not serve the same purposes as uplands, but can often be made to do so by drainage or filling. Thus, there is a paradox: while

10. This was the conclusion of the National Research Council Committee on Characterization of Wetlands (National Research Council 1995). Many observers of wetland regulation still cling to the view that the Corps is not sufficiently enthusiastic about pushing the regulations. The history of wetland regulation since 1972 suggests, however, that the effectiveness of the regulations is limited more by the ability of the public to resist them, both legally and otherwise, than by the nature of the agency that is charged with enforcing them.

wetlands seem like land to their private owners, they also are part of the surface water system that is protected by the Clean Water Act.

Because the American concept of private property is closely associated with ownership of land, the protection of wetlands seems to challenge the traditional rights of property owners, and in this sense differs from most other kinds of environmental regulation.[11] When a government removes an individual property right without just compensation, it is said to be guilty of *taking*.[12] Thus, the regulation of wetlands raises a takings issue.

Takings issues are resolved through courts. As of year 2000, legal challenges to the Clean Water Act have upheld broad application of the act to all connected surface waters, including wetlands on private property, and without direct compensation to landowners.[13]

Hypothetically, a regulatory system, through flexibility of permitting, could reduce the likelihood of regulatory takings and at the same time sustain the goals of the Clean Water Act. For example, if a parcel of land has high value to an individual or business but is of low value in supporting the goals of the Clean Water Act, a permit might be granted for its destruction. Thus, the protection of wetlands could

11. Property rights are constrained by a wide variety of factors such as creation of nuisance (e.g., support of invasive weeds on farmlands) or various forms of eminent domain. The weight of most environmental laws, however, falls not on individual property owners or on land but rather on public or private utilities or corporate entities in the form of restrictions on the release of pollutants to air and water.

12. Implicitly, this means *unconstitutional taking*; government can take legally under certain conditions. Reference is to the 5th Amendment of the Constitution: "nor shall private property be taken for public use without just compensation."

13. One legal test is whether or not the property holder has been denied all uses of the property, or merely a preferred use. When the property retains some usefulness to the owner, courts are likely to deny the occurrence of regulatory takings, even though the owner may be frustrated in using the property as desired. A second test is the degree to which a given use is consistent with societal expectations or values. A recent case that caused a stir in the legal community is *Lucas v. S. Carolina Coastal Council* (1992). Lucas proposed development of beachfront property adjacent to other such property but was denied permission to do so by regulatory authority, and successfully challenged the regulatory decision in court. In this case, societal interest, represented by the regulatory authority, could not claim that this particular beachfront commanded so much respect that private owners, by plan or custom, had been denied the right to build upon it; the prohibition against building was adopted after Lucas had bought the property. Some legal scholars view the *Lucas* decision as having implications for wetland regulations, while others find it consistent with past judicial support of wetland regulations from the viewpoint of takings. A good discussion of these matters can be found in McElfish (1994).

be selective rather than universal. In practice, there is considerable flexibility in the regulatory system that protects wetlands (as discussed below); permits are granted by the thousands. The puzzle is how this flexibility fits into the legal framework of wetlands protection and into national policy, which appears to be founded on the premise that the total amount of wetland in the United States will either stay the same or increase.

Three Little Words

Because the destruction of wetlands is not categorically prohibited, the Army Corps of Engineers has a fair amount of discretion in issuing permits. At the same time, the Clean Water Act does not instruct the Corps to make value judgments on wetlands and issue permits accordingly. The Corps does use judgment in issuing permits, but with minimal legal guidance and to an extent that is limited by the possibility of challenge by the EPA in its capacity as wetland watchdog, or by parties external to the government.

The executive branch, which presumably is the oracle of national policy, has declared under both parties, through the very lips of the President in each case, that the policy of the United States is *no net loss* of wetlands.[14] Probably any appealing phrase of fewer than four words has some chance of becoming a national policy, but one would hope that some thought went into presidential endorsements of *no net loss*.

In a speech before the National Wildlife Federation in 1993, Mr. Edwin H. Clark II, an authority on societal aspects of environmental regulations, described the birth and development of the *no-net-loss* policy (Clark 1993). Mr. Clark recalled that the controversy over protection of wetlands had prompted the EPA in 1987 to convene a group called the National Wetlands Policy Forum for a discussion of national goals in relation to wetlands. The forum was diversely constituted; it included individuals of both centrist and polar positions. One product of the forum was a policy subsequently described as *no net loss*.[15]

14. Presidents Bush and Clinton both explicitly endorsed the policy of *no net loss*. Both administrations, however, also endorsed initiatives that could be considered antithetical to the policy of *no net loss*.

15. The findings of the forum were presented at length in report form (Conservation Foundation 1988). Obviously, the forum members did not endorse the use of a three-word descriptor to summarize their entire report.

It seems amazing that a group as diverse as the forum could distill its opinion into such a succinct and seemingly clear policy directive as *no net loss*. Mr. Clark, who was a member of the forum, explained how this happened. First, there were many footnotes and caveats, all of which were discarded by a media, a government, and a public hungry for some decisive, understandable statement, as *no net loss* appears to be. In what could be a case study for plasticity of the human mind, Mr. Clark described the variety of interpretations that were attached by various forum members to each of the three little words. Beginning with "no," and referencing a well-known country song ("What part of 'no' don't you understand?"),[16] Clark noted that various forum members attributed diverse interpretations to each of the three little words. For example, "no" can be absolute or relative, depending on one's upbringing. "Net" is a tricky word because it suggests that there could be losses if the losses are offset by gains. In fact, individuals who have strong interests in draining or filling wetlands for various purposes support the notion that wetlands lost in one location can be replaced by wetlands created at another location that is more convenient or less valuable. This replacement philosophy raises unanswered questions about the feasibility of creating wetlands from scratch. Even the word "loss" presents problems. For example, do we mean loss of area or loss of function? If all of the functions of a wetland can be preserved, can its area be reduced? Also, are we evaluating loss over the short term or over the long term, and are we committing to the creation of new wetlands or merely to the preservation of existing ones? And so on.

It is clear from Clark's analysis why presidents of very different political persuasions have been willing to endorse a policy of *no net loss*: under creative management, such a policy can mean just about anything. While we seem to have a national policy, we do not have a binding interpretation of it.

There are no universally accepted figures on recent changes in wetland acreage for the entire nation. Even so, it seems likely that the total acreage of wetlands is decreasing,[17] although more slowly than in the past. The decrease is understandable, given that the permitting system

---

16. Lorrie Morgan, *What Part of No*, BNA Records 07863-66047-2/4 (1992).

17. The contiguous United States between 1986 and 1997 lost about 58,500 acres per year, or about 20% of the loss rate in the previous decade, according to the U.S. Fish and Wildlife Service (Dahl 2000).

does not cover all wetlands and because the Army Corps has issued tens of thousands of permits for wetland conversion. Most permits involve small areas, and many are offset,[18] at least in theory, by creation of new wetlands (*mitigation*); however, some slippage is inevitable. Thus, *de facto*, the national policy seems to be *slow net loss*.

Exclusions, Exceptions, and Excuses

Comprehensive regulations are usually greased with a few loopholes, as is Section 404. Excluded from full coverage are wetlands that cannot be classified as waters of the United States; these wetlands do not fall under the jurisdiction of the Clean Water Act. Exceptions are specified in general terms by the text of the Clean Water Act and in more specific terms by general permits covering entire classes of sites or activities. Even excuses are recognized administratively through an *after-the-fact* permit program (see below).

As explained above, the Clean Water Act applies only to waters that have a Commerce Clause connection. The main connection occurs through navigability, which has been given sufficiently expansive interpretation to include all surface waters that are not classifiable as isolated. Even isolated waters can have a Commerce Clause connection through migratory waterfowl or through recreation associated with foreign or interstate commerce; however, the degree of protection

18. Permitting can be summarized by the number of permits in various categories. In fiscal year 1994, there were 48,000 Section 404 permit decisions. Almost 40,000 of these passed through the general permitting system (i.e., they were permitted as part of a class of activities), about 3800 were handled by issuance of individual permits, about 4200 were withdrawn, 358 were denied, and a few hundred had other miscellaneous fates (U.S. Army Corps of Engineers, personal communication, 1995). In 1995, the Corps received 62,000 requests for permits and permitted conversion of 25,000 acres of wetland, but also required creation of 45,000 new or restored acres of wetland as mitigation (M. Davis, oral presentation, May 29, 1996, Washington, DC, Environmental Law Institute workshop). Thus, it would appear that the Corps program produced a net gain of wetland acreage in 1995, but it is not clear how much acreage escaped permitting, how successfully the mitigation acreage actually substituted for the permitted acreage, and how many thousands of small projects passing through the general permitting system did not appear in the acreage statistics. Without characterizing the Corps program as ineffective, it is still difficult to believe that the nation has avoided net loss of wetlands given all the possible ways in which wetlands can slip through the regulatory net (Babcock 1991). Some current statistics appear on the Army Corps of Engineers web site (http://www.USACE.army.mil/inet/functions/cw/cecwo/reg/).

under the Clean Water Act is less for isolated waters than for waters that fall under the expanded definition of navigability through their direct connection to waters that literally are navigable. Nevertheless, it is important to remember that federal protection of wetlands applies only to wetlands that have a Commerce Clause connection, which by present interpretation includes the majority of wetlands, but not all wetlands.

The Army Corps issues general permits that cover categories of activities or categories of sites. Conversion of wetlands under these permits is possible without a site-specific permit unless Army Corps or state or regional governments overrule the applicability of the general permit.

General permits fall under three major headings: regional, programmatic, and nationwide. The regional general permits cover specified small-scale activities within a particular geographic region. Programmatic permits are a means by which the Army Corps defers to state, local, or tribal regulatory programs that are consistent with the federal program. The nationwide permits are the most often used and most significant of the general permits.

Section 404 contains general language that offers exceptions for small-scale activities not having, individually or collectively, a significant adverse effect on wetlands. Thus, a small backyard project involving the elimination of some small bit of wetland, unless prohibited by state or local government, has been possible without a permit. Section 404 also explicitly excludes normal agricultural practice such as maintenance of existing ditches on farms. There is no information on the actual or potential loss of wetlands under these generalized exclusions from permitting requirements.

Nationwide permits are a bit of administrative genius that avoids head-on collisions between Section 404 and small-scale activities that individually affect wetlands in only minor ways. Examples include installation of docks, road crossings, outfall structures, etc.[19] While the purist might object that these activities are intolerable if they conflict in any way with the protection of wetlands, most people would see the need for such activities to proceed without obstruction if they do not involve large areas or cumulative major damage to wetlands.

19. See www.USACE.army.mil for descriptions of the more than 40 nationwide permits now valid. Recent changes mostly involve filling the void left by suspension of NWP 26, the most controversial of the nationwide permits.

The most broadly applicable nationwide permit until 1997 was Nationwide 26. Unlike the other nationwide permits, 26 did not apply to a specific kind of activity such as installation of docks. Instead, it loosened the permitting requirements on the basis of area for non-tidal wetlands that are "isolated" or classifiable as "headwaters" (i.e., below 5 cubic feet per second mean flow). Until 1997, projects in such areas involving less than 1 acre of wetland could be converted without a site-specific permit and without notification of the Army Corps. Conversions smaller than 10 acres could be accomplished without a site-specific permit following notification (including maps) of the Corps. Many small projects passed through this loophole. The existence of an area-based loophole, of course, attracted considerable attention. One regulatory staffer in Florida remarked that the number of past Florida development projects totaling 9.9 acres would seem quite a coincidence to anyone not familiar with Nationwide 26.[20]

In 1997, Nationwide 26 was placed in interim status under new thresholds: up to 0.3 acres for no prior notice (but notice after the fact was added for the purpose of establishing some record of cumulative change) and 0.3–3.0 acres for prior notification. Thus, the regulations have appeared much more restrictive after 1997, but critics have claimed that the change would have little effect on loss (Finder and Reiness 1997). One beneficial possibility is that the loss is better documented under these new regulations. Issues surrounding the former NWP 26 are still in flux.

The reasons behind nationwide permits, including Nationwide 26, are several. First, the relative hardship of a permitting requirement for a small business or individual dealing with a small amount of property is disproportionately high, and loopholes offset this burden. Second, and possibly even more important, is that the caseload for review by the Army Corps would be staggering and well beyond the capabilities of its regulatory staff if projects of all sizes were scrutinized with equal intensity. Third, the general degree of aggravation caused by wetland regulations to society as a whole is minimized if attention is focused on larger projects, and at least over the short term, the bulk of

20. The most egregious abuse of Nationwide 26 occurred through the cynical formulation of small projects that in fact were part of a large development plan. The Corps and the EPA have effectively opposed this abuse in most places where it was occurring (e.g., parts of Florida). The language of Nationwide 26 was also changed legislatively in 1991 to prevent this type of abuse (*Federal Register* 56:59110 [1991]).

wetlands may be protected by this means. Nevertheless, it is difficult to see how a policy of *no net loss* is consistent with a constant stream of small wetland conversions passing through the national protection system. This may be a case of national self-deception.

Violation of the Clean Water Act is punishable by severe penalties, but excuses are accepted in many cases in lieu of penalties. The means by which the Army Corps accepts excuses is the after-the-fact permitting program, which involves a cease and desist order for the violator followed by review of the action by the Army Corps, which will likely require modification of the project to reduce overall damage to wetlands. The work is then completed under permit, even though it began improperly without a permit.

Food, Security, and Wetlands

One of the least desirable jobs in the federal government must be explaining wetland regulations to farmers. One can imagine schooling the farmer on different types of land and on the values of wetlands. If one ducked the first punch, one could proceed to explain the virtues of sustaining biodiversity and water quality through preservation of wetlands for the good of society at large. One might save for the very last moment, perhaps to be shouted through a car window, an assertion of federal authority over wetlands on farms.

Wetland regulations can be economically harmful to farmers, but economic harm probably is not the main reason for agrarian opposition to wetland regulations. Farmers often are offended well beyond their economic losses.[21] This reaction comes from their sense of ex-

---

21. Personal observation. Meetings about wetlands in farming regions typically attract large numbers of landowners. Outrage is a common sentiment; it is supported by a string of anecdotes involving regulatory action that seems stupid or perverse. The underlying problem is that Section 404 cannot be applied to farmlands without effects on customary agricultural practice and the efficiency that comes with a farmer's sole authority over the land. In fact, Section 404 of the Clean Water Act exempts normal farming, silvicultural, and ranching activities. This exemption would cover plowing of lands customarily plowed, maintenance of existing ditches, etc., but does not extend to the conversion of wetland to cropland. As attested by massive conversion of wetland to cultivated land since colonial times, such conversion has been normal in agriculture by any common-sense interpretation of historical practice, but now is prohibited by the federal government except under permit.

clusive right to manage the property that they own and from their pride in past management of the land. Some also are insulted by the implication that they have been unable or unwilling to protect and manage their wetlands in a reasonable way, or that they would be unresponsive to the priorities of society around them.

The U. S. Department of Agriculture (USDA) often has served as a buffer between the agricultural community and the rest of society; Congress and the executive branch frequently have sought refuge in this buffer. The outpouring of objections from agriculturalists about wetland regulations led in the mid-1980s, through the USDA, to congressional separation of regulatory practice on agricultural lands from regulatory practice in general.

A watershed year for wetland regulation was 1985, when Congress passed the 1985 Food Security Act (1985 FSA), which contained intended solutions to problems that had plagued the application of wetland regulations to agricultural lands. Main features of the 1985 FSA and its subsequent amendments include: (1) definition of wetlands, (2) definition of agriculturally used or altered wetlands, and (3) establishment of punishment for alteration of wetlands ("swampbusting") after 1985.

Even though regulation of wetlands had been legislatively authorized since 1972, wetlands were never defined in Clean Water Act legislation. Thus, the 1985 FSA broke new ground by defining wetlands legislatively for the first time. The significance of the definition is explained in chapter 2.

The drafters of the 1985 FSA had very little choice about defining wetlands because one of the main purposes of the bill was to define classes of agriculturally altered wetlands; the task would have proven very awkward without a general definition of wetlands. With the definition of wetlands as a cornerstone, the 1985 FSA proceeded to define and describe three categories of wetland affected by agriculture: (a) converted wetlands, (b) prior converted cropland, and (c) farmed wetlands.[22] The definitions were supported with an identification manual (FSA manual).

Converted wetlands, according to the 1985 FSA, are wetlands that

22. The terms are explained by D. Snyder (1995). The terminology is confusing and must be committed to memory rather than taken at face value. "Converted wetlands" could have been called *wetland converted to cropland after 1985*, and "prior converted cropland" could have been called *wetland converted to cropland prior to 1985*.

were drained or otherwise altered after the threshold date set by the 1985 FSA (December 23, 1985) for the purpose of supporting agricultural production. The creation of converted wetlands by agricultural practice was prohibited by the 1985 FSA, with certain exceptions and exclusions. Prior converted cropland, which is sometimes (and more properly) referred to as prior converted wetland, includes all wetlands converted to cropland prior to the threshold date of the 1985 FSA. Maintenance of drainage and continued agricultural use of prior converted cropland is specifically approved by the 1985 FSA.

While it was probably never intended that wetlands already drained for agricultural purposes would be forcibly restored through the Clean Water Act, farmers were understandably nervous that this could prove to be the case and that they might therefore need to abandon the use of lands already in production. In addition, farmers wanted assurance that they could maintain drainage on lands already drained. The identification of prior converted croplands accomplished these purposes by exempting from regulation any wetlands that had been converted to agricultural use prior to the 1985 FSA threshold date.

The definition of farmed wetlands is more subtle and affects much smaller acreage but is of great practical importance in some regions. Farmed wetlands have the hydrologic characteristics of wetlands (i.e., they are not drained) and the soils of wetlands but are subject to practices designed to produce a crop (e.g., growth of hay in a wet meadow). Under the 1985 FSA, continued farming of farmed wetlands is allowed, provided that it is not accompanied by new efforts at drainage or filling.

Prohibitions against new conversion of wetlands to agricultural use after 1985 as specified by the 1985 FSA would have seemed ludicrous without some sort of penalty. For perspective, penalties for violating the wetland protection provisions of the Clean Water Act are of a criminal nature, even though errors of intent or ignorance typically are remedied through mitigation requirements involving the use of money for the restoration or creation of wetlands intended to replace altered wetlands.

The 1985 FSA contains a provision designated as "swampbuster," the coinage for which presumably makes the penalties seem more like down-home remedies than the administrative sanctions that they truly are. Swampbusting, which is the destruction or alteration of wetlands on agricultural lands contrary to the provisions of the 1985 FSA, is punishable by withdrawal of certain agricultural subsidies.[23] The penalties may seem minor by comparison with the penalties associated with the Clean Water Act, but must be evaluated in context; within a relatively short span of time, the role of the USDA had changed from facilitator of wetland conversion to guardian of wetlands on agricultural lands. It is surprising that the resulting functional transition could occur at all, and therefore not surprising that it has occurred slowly.

Hindsight on the 1985 FSA has shown that the bill was unnecessarily focused on restrictions. Subsequent amendments have added federal incentives through the use of reserve programs for agricultural wetlands.

Between 1985 and 1994, there was considerable confusion over the hierarchy of responsibility for mapping of wetlands on agricultural lands. While the USDA Natural Resource Conservation Service (NRCS) was charged with doing the mapping, which was performed according to the FSA manual, the requirements of the Clean Water Act, as interpreted independently by the Army Corps, also were applicable. This double jeopardy system drove farmers wild, and the message eventually reached Washington in sufficiently potent form that the Clinton administration, through a memorandum of agreement (EPA 1994), declared NRCS the preeminent mapper on agricultural lands and left the Army Corps with the lead responsibility for all other lands. Permitting still is vested with the Army Corps, but the Corps is expected to honor NRCS determinations unless it can make a case that the determinations are flawed. As before, the EPA stands in review capacity over all permitting and the U.S. Fish and Wildlife Service provides technical comment and expertise.

Federal protection of wetlands was initially based on the assumption that wetlands are natural landscape features that can be defined objectively. Critics of the Clean Water Act often have argued, however, that wetlands have been defined arbitrarily. If these critics were

---

23. The details are complex; they were modified in 1990 and again in 1996. The modifications seem to have been motivated in part by a desire to clear up misunderstandings and misapplications of the penalties, and in part by a desire to soften the effect of the penalties. For example, the 1996 Farm Bill excluded crop-insurance eligibility from the list of penalties that can be imposed under the 1985 FSA and also gave the secretary of agriculture much broader authority to waive penalties in individual cases. The most recent legislation emphasizes incentives (e.g., conservation reserve and wetlands reserve programs) rather than punishment.

correct, politically expedient modification of the regulatory system might seem defensible. Otherwise, as explained in chapter 2, the use of dual regulatory systems for agricultural and nonagricultural lands is undesirable because it sets the stage for arbitrary differences in the kind and amount of protection that wetlands will receive.

Manual Dexterity

With the extension of the Clean Water Act's jurisdiction to wetlands, the Army Corps of Engineers became guardian of some 100-million acres of swamp, marsh, and bog in the contiguous states and another 170-million acres in Alaska. Even though the Corps has never shown itself to be intimidated by any project of earthly proportions, the prospect of eternal vigilance and endless permitting on lands of such extent and dispersion must have been daunting.

The Army Corps initially delegated permitting responsibilities to its district offices, which composed regional guidelines for identification and mapping of wetlands and the issuance of permits. The identification of wetlands became known as *determination* and mapping of wetlands became known as *delineation*, which sounds every bit as bureaucratic as it can be in practice.

Delegation of regulatory conventions to districts eased the Corps into permitting through recognition of regional variation in kinds and amounts of wetlands, economic and societal conflicts with the protection of wetlands, and agricultural practice. Independence of regions, however, invites disparities in protection. Thus, the Army Corps of Engineers created a guidance manual (U.S. Army Corps of Engineers 1987; *USACE Wetlands Delineation Manual*, or 1987 Corps manual). This manual, which subsequently was supplemented with guidance documents, prescribed the methodology to be used in identifying and mapping wetlands throughout the nation.

The centerpiece of the 1987 Corps manual was the so-called *three-parameter approach*. This unfortunate choice of terminology, which conjures up the use of differential equations and mainframe computers, is intended simply to mean that the system for identification and mapping of wetlands is based upon three factors: water, soil, and vegetation. Thus, the 1987 Corps manual can be described either as *a three-parameter approach to the delineation of wetlands*, or, equally correctly, as *the use of soil, water, and vegetation to map wetlands*.

While the Corps was preparing its 1987 manual, Congress passed the 1985 Food Security Act. The act required the NRCS to map wetlands on agricultural lands and to develop and use its own manual (since revised as the National Food Security Act Manual 1994), which it prepared and issued prior to the completion of the 1987 Corps manual.

Real confusion began when the federal agencies prepared an interagency manual, which was released in 1989 (1989 interagency manual). Unlike the 1987 Corps manual, which was primarily the product of the U.S. Army Corps of Engineers, the 1989 interagency manual was a cooperative effort involving not only the Army Corps but also the EPA and the U.S. Fish and Wildlife Service. It was intended to be better than the 1987 Corps manual in a technical sense because it was more thorough in applying scientific principles to the mapping of wetlands and reflected the considerable experience that had accumulated through application of the 1987 Corps manual.

The 1989 interagency manual was lashed by a storm of protest. Many saw it as an attempt to broaden the legal boundaries of wetlands in an unreasonable way. This impression came especially from some early field applications of the new manual, which gave much more extensive boundaries for some wetlands than the 1987 Corps manual. Many landowners, agriculturalists, and developers began to feel that the federal government might be using wetland regulation as an excuse for broad incursion of federal authority onto private property.

Some of the early tests of the 1989 interagency manual were probably flawed, or at least misleading (R. Theriot, personal communication). In addition, even a fairly specific manual leaves room for interpretation, much of which can be standardized by the use of guidance letters and other supplementary documents. Thus, a manual that appears too tight can be loosened somewhat by regulatory convention; the 1989 interagency manual could have been adjusted in this way. These reasonable points, however, were lost in the fierce opposition to the manual. The 1989 interagency manual became a symbol of all that could possibly be wrong with wetland regulation; it was shouted down without ever becoming the standard for federal regulatory practice. Permits continued to follow the 1987 Corps manual.

The Bush administration was sympathetic to continuing protests over regulation of wetlands and supported the preparation of a set of revisions (1991 proposed revisions) to the 1987 Corps manual. The revisions contained some technical improvements, but fundamentally had the effect, which was probably intentional, of arbitrarily shrinking the legal boundaries of wetlands by requiring certain hydrologic evidence that normally is not available. Proving the existence of wetlands under the 1991 proposed revisions would in many cases have been somewhat akin to proving murder without a corpse or a weapon: possible but not likely.

Defenders of wetlands nullified the 1991 proposed revisions. The Army Corps continued to use the 1987 Corps manual and the NRCS continued using the 1985 FSA manual.

The tennis match over manuals left many nonexperts with a growing suspicion that the definition of wetlands must be more a matter of politics than of science. Thus, Congress asked the EPA to convene, through the National Academy of Sciences, a committee that would survey the field of battle and make recommendations. The academy's response, which was issued in 1995 as a report through the National Research Council (NRC), was reassuring to centrists in its conclusion that regulatory practice as established by the 1987 Corps manual, the 1985 FSA manual, and the numerous guidance letters accompanying these documents, had been fundamentally sound (National Research Council 1995). In a more controversial mode, the report confronted the thorny problem of agricultural/nonagricultural duality in defining and mapping wetlands by recommending that all wetlands be identified and mapped by use of a common manual and a common definition. The NRC report also found great fault with Nationwide Permit 26 and recommended several changes in methodology, as well as a general upgrade in understanding of the connections among the three factors by which wetlands are identified and mapped. Most importantly, the NRC report endorsed the view that wetlands are units of nature that can be identified by objective means capable of producing repeatable results.

2

WHAT A WETLAND IS, AND ISN'T

Uneven distribution of water causes the landscape to be divided into four parts: lakes, streams, uplands, and wetlands. Lakes are distinguished from the rest by the presence of water standing at considerable depth over long intervals. As a rule of thumb, the water of a lake is deep enough to prevent the growth of rooted vegetation over most of its area. Streams (and rivers, which are simply large streams) are moving waters confined by a channel. Uplands are seldom or never covered with water and are fully saturated to points near the surface only during cold weather or for short intervals during the growing season. Parts of the landscape that remain after the exclusion of lakes, streams, and uplands are wetlands.

Wetlands may be inundated constantly, seasonally, or never. If never inundated or inundated seasonally—as are the lands on river floodplains—wetlands can be dry at the surface. Even so, they will be very wet just underneath for all or much of the growing season. Wetlands thus are places where heavy equipment is likely to churn up mud even after the weather has been dry for some time.[1]

1. This agro-mechanical definition of wetlands will not be found in federal manuals but works pretty well because, as shown in later chapters, wetlands are saturated within at least one foot of the surface for extended intervals. Thus, dry transgression of wetlands by bulldozer, dragline, or even most farm tractors usually is not possible during the growing season. The genesis of the definition is an exchange between the author and Senator Lauch Faircloth (R–NC) during 1995 hearings related to HR 961. Senator Faircloth had asked for a practical definition that could be given to a farmer who might

Water provokes important responses from plants, animals, and microbes that in turn cause uplands, wetlands, lakes, and streams to be fundamentally different from each other. Where water stands for long periods at a depth of several feet, we find organisms with life cycles that presume the constant presence of water, and the sediment lacks the properties of a true soil. Where the land is characteristically dry at and near the surface during the growing season, the assortment of species is different, and typically there is a soil with well-defined structure in the form of layers (horizons). Wetlands are different from either of these in that their resident communities can tolerate water saturation or inundation for long intervals but usually can withstand desiccation of the surface as well. They form soils, but of a class distinct from the soils of uplands. Thus, the uneven distribution of water causes some predictable kinds of variation in living communities and soils, both of which can be used in defining and mapping wetlands.

The *E* Word

In the early days of their history, a number of the midwestern states formed organizations called natural history surveys. The natural resources of the United States were then so unfamiliar that the states needed an accounting of them. Formation of the surveys led to the employment of scientists who were charged with inventorying natural resources with an eye toward commerce and development.

One of the early employees of the Natural History Survey in Illinois was Stephen Forbes, a zoologist with interests in aquatic environments. Forbes and his coworkers sampled the fishes and their aquatic foods in the rich overflowed swamplands along the Illinois River. Using his field experience and knowledge of aquatic organisms and their ways of living, Forbes wrote an essay on the architecture of the environment (Forbes 1887). We now regard this essay as a classic because it proposed a new way of viewing and analyzing nature.

Forbes's essay dealt specifically with lakes, although his ideas proved transferable to other kinds of environments. According to Forbes, a lake can be viewed as a unit in the hierarchy of nature rather than a mere collection of animate and inanimate objects that happen to

---

not have time for graduate studies in wetland science. In the pressure of the moment, tractors came to mind, and thus the definition.

be located in the same place. He argued that the lake is a microcosm, or little world, in which the components are knit together to make a system that, when viewed as a whole, has properties that could not easily be imagined from a list of components.

Forbes pointed out that the lacustrine microcosm is much like an organism. It has, for example, a metabolism that includes both synthesis (photosynthesis and organismic growth) and degradation (respiration and decomposition). A lake also sustains internal flows of critical materials, which we now refer to collectively as nutrients. A lake has a definite internal organization in the form of physical or chemical gradients that cause gradients in the distribution and abundance of organisms. Finally, the organisms of the lake constantly affect each other through a network of interactions involving competition and predation. In this way, Forbes not only identified the advantages of studying nature above the level of individual species or individual organisms, but also forecast with remarkable accuracy the most profitable ways of approaching the analysis of large, integrated units of nature.

Large, integrated units of nature, such as Forbes's lacustrine microcosm, are today referred to as *ecosystems*. This word, which is essentially shorthand for *ecological system*, was coined by A. G. Tansley (1935), a plant ecologist, almost 50 years after Forbes published his essay. Forbes's concept of ecosystems now defines one of the half dozen major branches of the ecological sciences.[2] It is here that we find much of the basic information supporting environmental management, protection, and maintenance of yield from nature. The science of ecosystems, however, remained out of public view for a very long time.

Ecosystem science might have been more widely advertised if it had not been burdened by holism and environmental politics. While Stephen Forbes was innovative in thinking holistically about the environment, holism now has been held in cheap embrace by so many people for so many reasons that it commands little respect. To vast numbers of scientists and engineers who must produce information and knowledge by the analysis of specifics, holism seems of no practical value, however comforting it may be to individuals who prefer

---

2. Some authorities attribute less significance to Forbes than would be suggested here. For an historical perspective, see McIntosh (1985). For a brief account of the other branches of ecology, see Lewis (1994).

not to trouble themselves with details. This mind-set overlooks, of course, the legitimate uses of holistic thinking in science and elsewhere. Understanding the U.S. economy, for example, requires not only knowledge about employment and interest rates, but also a holistic view of the way in which components interact to produce otherwise unpredictable outcomes. Thus, we cannot discount ecosystem science simply because it is fundamentally holistic.

The baggage of environmental politics may have been even heavier. While ecosystem science grew rapidly in scope and sophistication after 1960 (Golley 1993), it became confused, as did the ecological sciences in general, with environmentalism. The *eco* prefix is sometimes presumed to be the creation of environmental extremists, or a thinly veiled attempt to propagate socialistic economic programs. The ecological sciences, however, could thrive in a political vacuum; their goal, which is to understand life in context with its surroundings, is no more inherently political than astrophysics.[3] Because more than one scientific thrust has been obstructed by congressional or public sensitivities, however, ecologists were advised as recently as the 1970s to avoid the use of the term *ecosystem* when approaching the federal bread box.

Remarkably, the stigma attached to the ecosystem concept has largely dissipated. Federal agencies have accepted *ecosystem management* as their guiding principle, although their public allegiance to this principle is accompanied by much private confusion over its meaning.[4] If this commitment is sustained, it could mean more realistic balancing of the multiple and often conflicting purposes for maintaining public lands. In the meantime, the *E* word can be used more freely, and its usefulness to society may become clearer.

The Ecosystem Quilt

The four components of a landscape dictated by the distribution of water (lake, stream, upland, and wetland) define four categories of ecosystems. Within each of these categories, there are numerous

---

3. Yet it is true that politics have influenced ecology and that ecologists have been political (Bocking 1997).

4. See Slocombe (1993). The confusion is not limited to administrators of federal agencies; ecosystem scientists also are still deciding how to respond (Christensen et al. 1996).

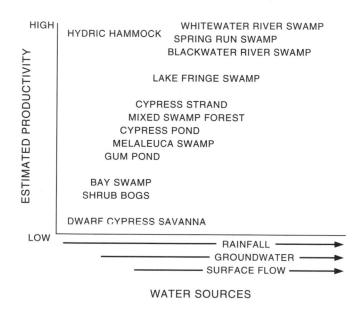

Figure 2-1. Illustration of the variety of wetlands that could be found within a single region. Wetlands of Florida are arranged according to productivity and water sources (redrawn from Ewel 1990).

classes and subclasses. For example, upland ecosystems can be divided into grasslands and forests. Similarly, among wetlands we can distinguish classes such as swamp, marsh, and bog. Classes can be further subdivided on the basis of hydrologic regimes, vegetation, or soil into subclasses that are often identified with specific geographic regions (Wilen and Tiner 1993). At the bottom of the hierarchy is the individual ecosystem. Thus, the landscape is a patchwork of ecosystems.

The regulation or protection of wetlands presents two kinds of challenges. First is to identify characteristics of wetlands that distinguish them from other kinds of ecosystems. The selection of these characteristics must take into account the possibility of considerable variation among different classes of wetlands, the existence of subclasses, and, even within a given subclass, of individual variation among wetlands (figure 2-1). The second challenge is to show how the boundary between a wetland and another kind of ecosystem can be located by means that are both objective and repeatable.

Solutions to the identity problem and the boundary problem require an explicit definition of wetland. In the absence of regulations, however, formal definitions would not be necessary. Consequently, wetland specialists did not put much energy into definitions until the identification of wetland became a legal issue. While attorneys scrutinize definitions down to their very prepositions and indefinite articles, scientists are almost indifferent to them. Thus, the Army Corps of Engineers, as it began to apply Section 404, could not simply adopt a standing scientific definition of wetland.

Three Definitions

A historian probably could find more than a dozen definitions of wetland, but at present only three definitions of wetland really matter in the United States: the 1979 USFWS definition, the 1977 Army Corps definition, and the 1985 FSA definition. Each of these reflects intensive debate influenced by fears that a sloppy definition would be impossible to apply or that an imprecise definition would promote either overextension or excessive contraction of the regulatory net that is intended for wetlands.

Wetlands by Way of Wildlife

The 1979 USFWS definition has deeper roots than either of the other two major definitions. The federal government began acquiring wetlands for the protection of waterfowl as early as the late nineteenth century; the agency mainly responsible for selecting these lands and advocating their purchase was the U.S. Fish and Wildlife Service. Consequently, the USFWS had an interest in the identification and inventory of wetlands in the United States long before the protection of wetlands through the Clean Water Act. In 1956, this interest led to the publication of USFWS Circular 39, which provided a definition of wetlands and a classification of wetlands that included 20 different wetland types.[5]

5. Shaw and Fredine (1956). The core of the 1956 definition is as follows: "lowlands covered with shallow and sometimes temporary or intermittent waters [including] shallow lakes and ponds, usually with emergent vegetation as a conspicuous feature, [but excluding] streams, reservoirs, and deep lakes [as well as] water areas so temporary as to have little or no effect on the development of moist soil vegetation."

The definition of wetlands in Circular 39 is relatively loose, as one might expect given its nonregulatory application. It refers to shallow or intermittent inundation as a distinction between wetlands and deepwater habitats (lakes), and it also mentions characteristic growth of vegetation that reflects prolonged saturation of the soil with water. A three-factor approach (water, soil, and vegetation) is implicit in this early definition.

The USFWS released a new definition in 1979.[6] By this time, the regulatory significance of definitions had become evident and the new definition reflected agency debate that the USFWS participants recall as having been agonizingly difficult. The 1979 USFWS definition, and particularly its treatment of the wetland boundary, identifies wetlands as having one or more of three attributes: (1) a predominance of plants specifically adapted to life in saturated soils (hydrophytes); (2) a predominance of soils showing evidence of development under saturated conditions (hydric soils); and (3) if the substrate is not a true soil (e.g., sand or gravel), saturation with water each year. The craftsmanship of this definition is evident from its phrasing. For example, the definition requires that hydrophytic vegetation or hydric soils be "predominant." This gives the interpreter some guidance: the presence of a few hydrophytes or a small patch of hydric soil does not define a wetland, nor will the presence of a few non-hydrophytes or a bit of uncharacteristic soil indicate that an area cannot be a wetland.

An important characteristic of the 1979 USFWS definition is its reference to any one of three factors, rather than to weight of evidence from a combination of factors. In evaluating this feature of the definition, one must recognize that the USFWS does not write permits or enforce regulations related to wetlands. Its function is to inventory wetlands, and the inventory function is facilitated by the use of single

---

6. Cowardin et al. (1979). The definition is as follows: "Wetlands are lands transitional between terrestrial and aquatic systems where the water table is usually at or near the surface or the land is covered by shallow water. For purposes of this classification wetlands must have one or more of the following three attributes: (1) at least periodically, the land supports predominantly hydrophytes; (2) the substrate is predominantly undrained hydric soil; and (3) the substrate is nonsoil and is saturated with water or covered by shallow water at some time during the growing season of each year." The publication also defined the boundary between wetland and upland as that lying "between land with predominantly hydrophytic cover and land with predominantly mesophytic or xerophytic cover; [or] the boundary between soil that is predominantly hydric and soil that is dominantly nonhydric; [or] in the case of wetlands without vegetation or soil, the boundary between land that is flooded or saturated at some time each year and land that is not."

factors, any one of which is usually adequate for identification of a wetland. A definition of this type reduces the practical difficulty of dealing with wetlands in which vegetation has been removed, soil analyses are not available, or the substrate is not a true soil. Despite its convenience for mapping, this feature of the USFWS definition has not been carried over to regulatory definitions because it appears to present unacceptably high likelihood for error in defining areas as wetlands that are not truly wetlands. This is a major concern for regulatory agencies, but less so for the USFWS.

Wetlands by Way of Water

The second definition of great importance is that of the U.S. Army Corps of Engineers, as adopted in 1977 (1977 Corps definition).[7] This definition, which is also accepted by the EPA, provides the main framework for enforcing the provisions of Section 404 of the Clean Water Act. The definition states that wetlands are, under normal conditions, subject to inundation or saturation of the surface sufficient to maintain a prevalence of hydrophytic vegetation. The concept of prevalence here parallels the concept of predominance in the 1979 USFWS definition. There is no specific mention of soil, even though soil is used in the 1987 Corps manual and is an integral part of the Corps' three-factor system involving water, soil, and vegetation. Water is mentioned, but not as an independent basis for defining wetlands; it is a condition required for the maintenance of hydrophytic vegetation. Thus, the entire definition centers around hydrophytic vegetation as an indicator of inundation or saturation. This reflects the particular value of hydrophytic vegetation in the identification and mapping of wetlands for everyday application of Section 404.

The reference to "normal circumstances" is the means by which the 1977 Corps definition recognizes wetlands from which the vegetation has been removed or altered. An area that can be expected to regenerate hydrophytic vegetation thus can be declared a wetland.

---

7. "Those areas that are inundated or saturated by surface or ground water at a frequency and duration sufficient to support, and that under normal circumstances do support, a prevalence of vegetation typically adapted for life in saturated soil conditions. Wetlands generally include swamps, marshes, bogs, and similar areas." (42 Federal Register 37, 125-26, 37128-29; July 19, 1977)

Wetlands by Way of Agriculture

A third important definition is that of the 1985 FSA manual (1985 FSA definition),[8] which specifies predominance of hydric soils and saturation or inundation sufficient to support hydrophytic vegetation under normal conditions. These two requirements are given as additive rather than alternative. The definition does allow, however, for the identification of wetlands in which vegetation has been altered. The first sentence of the definition also excludes wetlands converted prior to the 1985 FSA, thus assuring farmers that wetlands drained before the threshold date for prohibition of drainage will not be considered wetlands even though they may have hydric soils as remnants of their earlier condition. The 1985 FSA definition does not allow the identification of wetlands anywhere that hydric soils are absent, which puts this definition in direct conflict with the 1979 USFWS definition. The significance of this difference will be explored in chapter 5.

The FSA definition also contains a significant postscript declaring that lands in Alaska cannot be wetlands if they have high potential for agricultural development and a predominance of permafrost soils. Here, the FSA definition illustrates the vulnerability of regulatory definitions to political considerations. While it seems perfectly legitimate for Congress to decide not to protect certain wetlands in Alaska (whether or not this is wise is another question), perversion of a basic wetland definition for this purpose is most unfortunate because it leads to confusion over what wetlands really are, and breaks the critical separation between technical points and political objectives. The Alaska provision is about as sensible as a definition of water referencing a covalent combination of two atoms of hydrogen and one of oxygen, except in Delaware.

8. "The term 'wetland,' except when such term is part of the term 'converted wetland' means land that—a) has a predominance of hydric soils; b) is inundated or saturated by surface or ground water at a frequency or duration sufficient to support a prevalence of hydrophytic vegetation typically adapted for life in saturated soil conditions; and c) under normal circumstances does support a prevalence of such vegetation. For purposes of this Act or any other act, this term shall not include lands in Alaska identified as having high potential for agricultural development which have a predominance of permafrost soils." (16 USC Section 801[a][16])

Weighing the Definitions

All three important definitions throw down the gauntlet of interpreta-tion for individual words and phrases. Some critical words can be in-terpreted by the application of common sense and conventional mean-ing, provided that the parties doing the interpreting are intent on being reasonable and objective. For example, the concept of normality can be interpreted variously, but reasonable interpretations are limited by common sense. More problematic are technical terms subject to defi-nition or interpretation without the constraint of broadly accepted usage. Two of the most troublesome terms in this regard are "hydrophytic vegetation" and "hydric soils." Both of these terms are essentially de-fined by federal committees. Because the two regulatory definitions (1977 Corps definition and 1985 FSA definition) of wetland are di-rectly pinned to these two terms, the conventions of the committees for defining them are critical to the definition of wetlands (as shown in chapters 4 and 5).

The NRC Committee that was formed in 1994 at the request of Congress to assess confusion over manuals inevitably had to deal with the matter of definitions. The committee considered skirting the issue by simply endorsing one of the existing definitions, but in the end con-cluded that the existing definitions, however valid, would be associ-ated with the missions or viewpoints of the agencies that created them. The committee also concluded that each of the three federal defini-tions has some shortcoming that could be corrected through a fresh start.

The NRC Committee proposed a reference definition for wet-lands.[9] The committee pointed out that regulatory definitions must always be weighed against some nonregulatory standard, simply be-cause regulatory definitions can be very far off the mark if they hap-pen to be overly influenced by political considerations (as illustrated

9. "A wetland is an ecosystem that depends on constant or recurrent, shallow inun-dation or saturation at or near the surface of the substrate. The minimum essential char-acteristics of a wetland are recurrent, sustained inundation or saturation at or near the surface and the presence of physical, chemical, and biological features reflective of re-current, sustained inundation or saturation. Common diagnostic features of wetlands are hydric soils and hydrophytic vegetation. These features will be present except where specific physicochemical, biotic, or anthropogenic factors have removed them or pre-vented their development." (National Research Council 1995) This definition is under consideration for formal endorsement by the Society of Wetland Scientists.

by the exclusion of extensive Alaskan wetlands from the 1985 FSA definition of wetland). Although the committee initially considered the creation of a reference definition to be a minor part of its work, it was drawn again and again to the definitional tar baby until virtually every word of the reference definition had been debated at length. The committee found the birthing of its 89-word reference definition at least as difficult as that of its accompanying 300-page technical analysis.

The NRC Committee's reference definition specifies: (1) that wetlands are ecosystems; (2) that they require in all cases recurrent and sustained saturation with water at or near the surface; and (3) that saturation must be sufficient to sustain distinctive physical, chemical, and biotic characteristics. The definition also states that the substrate of wetlands will typically consist of a hydric soil and that the biota of a wetland will typically include hydrophytic vegetation.

The ecosystem frame of reference for the NRC definition exposes the impossibility of treating the water, substrate, and organisms of wetlands as disconnected entities, either for purposes of identification and regulation or for management and protection. The ecosystem label warns us that tampering with any key attribute of a wetland is likely to have a ripple effect on other attributes, much in the way that tampering with an engine could produce a variety of changes in its performance. The ecosystem perspective also encourages us to look for cause and effect relationships that will facilitate the mapping of wetlands and the preservation of their distinctive characteristics.

The NRC Committee also wanted to emphasize the special importance of water in creating and maintaining wetlands. Water is not simply one of three factors that characterize wetlands. Water, in the form of particular kinds of hydrologic conditions, is the specific cause of wetlands; the characteristic substrates and organisms of wetlands develop in response to water. For this reason, a wetland can regenerate itself if the organisms are removed, or even if the substrate is removed and replaced with an uncharacteristic substrate, but not if the water is removed. In fact, where water has been removed, the substrate (typically a hydric soil) continues for a while to be characteristic of a wetland, as do the organisms (usually by the presence of long-lived hydrophytes); but the wetland ecosystem itself is dead, and the direction of change, which will soon occur in the vegetation and eventually in the soil as well, is toward upland.

The NRC Committee wanted its definition to be as general as possible without encompassing ecosystems that are not wetlands. For this reason, the committee specified that wetlands will show physical, chemical, and biological evidence of recurrent and prolonged saturation with water, but it did not limit this evidence to hydric soils and hydrophytes. In principle, there is no compelling reason why wetlands cannot develop where hydric soils do not form or have not had time to form, or where hydrophytic vegetation cannot grow for some reason having to do with factors other than saturation of the substrate with water. For example, a playa in the western United States may contain sufficient salt to kill hydrophytic vegetation, but would still develop aquatic invertebrates and store the resting stages of these invertebrates over dry intervals. A sandy substrate continuously saturated by moving water may never become deoxygenated, and thus would be outside a formal definition of hydric soil, but could still support a water-adapted community having all of the essential features of a wetland. At the same time, the committee wanted to emphasize that most wetlands are associated with hydric soils (see chapter 5) and hydrophytic vegetation (see chapter 4).

The reference definition of the NRC Committee may not be the only possible definition that could serve as a standard for evaluation of regulatory definitions, but it does carry the strength of consensus from specialists of widely differing viewpoints and backgrounds and does not reflect any mission to preserve, protect, exploit, or eliminate wetlands. As future regulatory definitions are proposed, they can be tested against the reference definition.

Wetland Taxonomies

Wetlands have been classified in many different ways.[10] Formal classifications often are detailed; they are quite useful to specialists but tend to give the more distant observer a headache. There are, however, two general and easy ways to classify wetlands. One is based on vegetation, and the other on physical features.

Most wetlands will fit reasonably well into a classification that divides wetlands into swamps, marshes, and bogs. This classification is based on vegetation: swamps are dominated by trees; marshes are

---

10. On December 17, 1996, the federal government adopted the system of Cowardin et al. (1979; note 6) as the federal standard for classifying wetlands.

dominated by grasses; and bogs are dominated by shrubby vegetation and often include large quantities of moss, particularly of the genus *Sphagnum*. Numerous differences among these three types accompany or are caused by the differences in vegetation. For example, the birds and mammals that occupy swamps may be substantially different from those that occupy marshes or bogs. Similarly, the productivity of bogs is typically lower than the productivity of swamps or marshes. Even though this classification separates wetlands well for some purposes, it does not reflect some of the important differences among wetlands. For example, wetlands of any of the three categories can be maintained by groundwater, by surface water, or by both. Thus, the three categories based on vegetation show substantial overlap in physical characteristics.

Wetlands also can be classified according to their physical characteristics. One such method that has many merits and has been in great favor with the Army Corps is the *hydrogeomorphic approach*.[11] According to this method, wetlands can be classified meaningfully on the basis of their water source and pattern of water delivery (hydro) and their physical conformation and position on the landscape (geomorphic). One major distinction on the hydro side is between wetlands that are maintained by surface waters and those that are maintained by groundwater or by precipitation. Another hydrologic distinction is among water flows that are vertical, horizontally unidirectional, and horizontally bidirectional. Under the geomorphic heading, there are distinctions among depressional and mounded wetlands, wetlands of river floodplains, and wetlands of estuary fringes.

Hydrogeomorphic classes tend to be functionally distinct in some ways. For example, wetlands receiving water only from precipitation would likely be very poor in nutrients, whereas those receiving groundwater would be slightly richer, and those receiving surface water would be richer still. Even so, a number of wetland characteristics are mixed across different hydrogeomorphic classes.

General classification schemes based on vegetation or physical factors break down for certain wetlands that are mosaics or mixed types.

11. The hydrogeomorphic classification system (HGM) is based on five geomorphic settings (river, depression, fringe, slope, and flat), three sources of water (precipitation, overland flow, and groundwater), and three hydrodynamic categories (vertical, horizontal unidirectional, and horizontal bidirectional). Combinations of these produce 45 HGM classes. Unaltered wetlands in a given class are used as benchmarks for altered wetlands of the same class (Brinson 1993, 1998).

As with any sort of classification, the user must be willing to accept the categories as a sort of platonic ideal from which most wetlands will differ some and some will differ much. Even so, these schemes can be useful in sorting out functional differences among wetlands or differences in the value of wetlands to humans.

Edges and Transitions

Wetlands lack sharp boundaries, which has caused some critics of delineation to argue that mapping is impossible, or at best arbitrary. It is true that wetlands fade into the adjoining upland or the adjoining lake or stream over a distance that may be as small as a meter or much in excess of 10 meters; this presents a boundary problem for mappers and is a source of controversy because of the economic consequences that may accompany placement of the boundary. Even so, there is no reason why boundary determinations cannot be repeatable and rational. Any one of several rational solutions is possible and the choice among these is a matter for policy rather than for science. For example, one could set the boundary at the outermost limit, where the last of the characteristic wetland features disappears, or one could take the opposite approach, which would result in a boundary line at the innermost part of the gradient that separates wetland from upland. Given that the practical concerns of defining wetlands too loosely or too restrictively both have merit, it might make sense to set the boundary between the innermost and outermost limits to the transition zone; in fact, this is the general result of most federal delineation practice.

Wetlands are sometimes viewed as transitions between dry land and open water. In fact, the 1979 USFWS definition characterizes wetlands as transitional environments. Some wetland scientists also favor this way of looking at wetlands. Ultimately, however, this may not be a very useful way to think of wetlands. For example, it is not true that wetlands are always located between land and water; they are often detached from lakes or streams. Furthermore, many of the most important characteristics of wetlands are not transitional at all. The biological productivity of wetlands is not in any sense transitional or intermediate between upland and open water; neither are the biological communities of wetlands merely blends of communities from uplands and open waters. Because wetlands are distinctive ecosystems, the idea of transition is of limited use in analyzing them.

3

WHAT WETLANDS DO, AND
HOW THEY DO IT

Those who question the wisdom of wetland regulations sometimes tell
a story about a landowner who proposes to build a fine home on a tract
of land that is largely wetland. This landowner lives in a region where
wetlands are abundant but is denied permission to build on grounds
that construction would involve filling a wetland. Because he owns
considerable property, the owner moves to higher ground, clears five
acres of mature upland timber, and builds his home quite legally in
this way. The irony is that the mature upland timber is much scarcer
locally than wetland, and the stupidity of the regulation is to have
forced someone to destroy the scarcer of two resources.

The names, places, and other particulars of this story vary with the
teller, but there is little doubt that wetland regulation has sometimes
caused an environmental loss greater than the value of the wetland that
is preserved. Perhaps the landowner in the story could have been given
an exemption had he only been allowed to argue the great value of mature upland forest in his particular region. As a practical matter, however, special pleading can defeat the intent of almost any regulation.
Thus, the rigidity in the regulation may be justified by its need to be
faithful to the general intent of the underlying law and not by a need to
be rational in every case. At any rate, the story may be specious as a
generalization in that the Army Corps in most cases would have
granted a permit to an individual for a small wetland conversion, or
the conversion would have been covered under a general permit for
small conversions (chapter 1).

The story does raise questions about justification for the special protection that we have given to wetlands. A skeptic may rightly ask what wetlands do that elevates them in regulatory priority above woodlands and grasslands. A related question is whether we need the 100-million acres of wetland that we now have in the contiguous 48 states, not to mention 170-million acres in Alaska, or if we might not be just as well off with 50 million or 5 million.

Functions, Values, and Justifications

People who evaluate wetlands for a living distinguish between functions and values of wetlands. To the wetland specialist, functions are all of the processes that occur in a wetland. For example, many wetlands capture sediment from surface waters; capture of sediment thus could be listed as a function. Functions also can be complex and even dependent on other functions. For example, maintenance of bird populations is a function of southeastern swamps but only through other functions that provide birds with habitat and food.

Values, in contrast to functions, are attributes about which humans have opinions. Values may be tangible, as would be the case for maintenance of water quality by wetlands, or intangible, as would be the case for aesthetic appeal of wetlands (Wilson and Carpenter 1999).

The dichotomy between functions and values is handy to wetland evaluators because it nicely separates things that easily become confused: values of wetlands are dictated by human opinion, whereas functions are not. A wetland will function as a sediment trap, regardless of the value that society may attach to sediment-free water. In contrast, values may change. For example, a large segment of society now attaches value to the biotic diversity that wetlands maintain, but might not have done so 50 years ago.[1] Value that can be measured

1. A national survey conducted by the U.S. Fish and Wildlife Service during 1996 showed that 87% of those surveyed find it personally important for the United States to maintain biological diversity. In progressively more-acid tests, however, the survey showed that 77% support the expenditure of more federal tax dollars for preservation of biodiversity but only 48% favor paying more federal taxes to achieve this end (the gap between the latter two numbers is the great conundrum of American environmental politics). Unpublished information from R. C. Pisapia, Assistant Regional Director North Region of the U.S. Fish and Wildlife Service (survey information from Belden and Russonello, Inc. 1996).

commercially may also change. Timber can be removed from wetland and sold, thus sustaining commercial value for the wetland,[2] but the value of timber may change with time.

Both functions and values differ from one wetland to another. This complicates the assessment of wetlands at a regional or national scale. For example, a marsh does not grow large trees and therefore lacks the commercial value associated with timber as well as the intangible value of birds that require large trees, but may have other values instead.

In sustaining Section 404 of the Clean Water Act, the United States has recognized that wetlands have special values. Wetland values, in turn, are sustained by wetland functions. Most people who support protection of wetlands are operating mainly on instinct, however, and not from specific knowledge of functions and values.

The functions and values of wetlands fall roughly under three headings: (1) flow of water, (2) quality of water, and (3) habitat for organisms. An overview of these will show how the protection of wetlands has been justified.

The Wetland and the Water Cycle

Water flows because of precipitation, without which it would pass to the lowest or coldest points and stop (mainly in the oceans, ice caps, and deep groundwater reservoirs). Evaporation and release of water vapor from plant leaves (transpiration) liberate water to the atmosphere as water vapor. Water vapor moves freely but, because there is a limit to the vapor-holding capacity of the atmosphere, water in the vapor state soon returns to earth as precipitation.[3] Thus, the general cyclicity of water is in principle neither profound nor complex, even though its analysis and prediction at specific times and places still consume the full attention of many brilliant minds.

In the hands of an expert, a picture of the water cycle can easily be-

2. The Tulloch Rule mentioned in chapter 1 (note 10) prohibits mechanized land clearing if such activity is accompanied by an addition or redeposit of dredged material in a wetland. In general, timber harvesting is allowed if it does not disturb the root system or involve the use of machinery that is prone to cause redeposition of excavated material.

3. The mean residence time of water in the atmosphere is approximately 9 days (Dooge 1984).

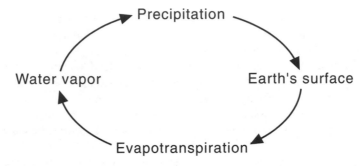

Figure 3-1. A simple view of the global water cycle.

come a thicket of arrows, boxes, and differential equations. At the opposite end of the spectrum is figure 3-1, a simplistic but adequate view
of the water cycle for present purposes.

Figure 3-1 does not indicate the existence of wetlands. This is
somewhat surprising given that Congress, in debating the protection
of wetlands, linked wetlands to the water cycle (chapter 1). Senator
Muskie, who may have been the first member of Congress to point out
that water "moves in hydrologic cycles," would have been more to the
point (but certainly less profound) if he had said instead that water
"moves downhill." It is not the water cycle itself that explains the significance of wetlands, but rather the downslope movement of precipitation (figure 3-2).

Water vapor can return to Earth at any elevation. If it falls above sea
level, as it usually will if it strikes land or inland waters, it will move
downhill toward the oceans,[4] which it will reach unless it first returns
to the atmosphere as vapor. All water flowing at the surface past any
given point on a stream or river is called the *runoff* from the watershed
above that point.

Precipitation may flow over the soil surface or beneath it. Overland
flow occurs only in the wettest weather. More often, runoff enters

4. There are two exceptions. In arid regions, water may enter valleys so dry that no
water leaves them except by evaporation. An example is the drainage basin of the Great
Salt Lake. A second exception is deep groundwater, which on a human timescale is ba-
sically immobile and, under some circumstances, can receive and store flow from shallow groundwater.

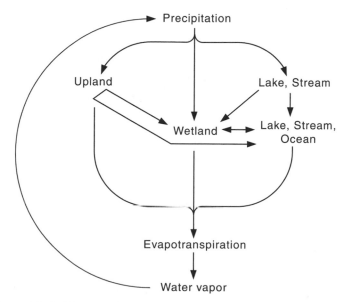

Figure 3-2. A diagrammatic view of the hydrologic basis for the significance of wetlands in the water cycle: Wetlands intercept waters en route to surface waters and also intercept waters that are already in transit within the drainage net.

streams or rivers in more subtle and orderly fashion by moving downslope either through a saturated soil or through a subterranean alluvial layer that is saturated with water. About half of the subterranean water derived from precipitation finds an exit at the surface within a few minutes to a few months, depending on circumstances. The rest is transpired by plants or evaporates if it moistens the soil surface, thus becoming water vapor by the combination of processes known as *evapotranspiration*. Subterranean water returns to the surface through seepage zones or discrete springs, and thus joins runoff.

Runoff, whether accumulated by seepage or the more occasional mechanism of overland storm flow, establishes a drainage net that reflects topography. The fringes of the drainage net, where land meets water, often are occupied by wetlands. Thus, in transit either underground or overland, water often passes through a wetland just prior to entering surface waters (figure 3-3).

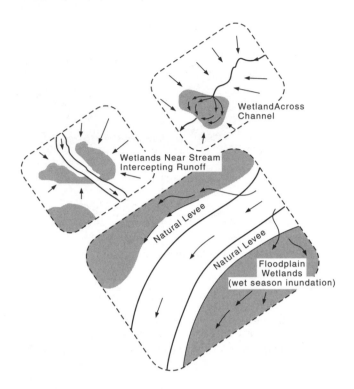

Figure 3-3. Diagrammatic illustration of the three main ways in which wetlands intercept surface water: Wetlands may lie across stream or river channels (common in small streams but possible elsewhere as well); wetlands may be a conduit to surface waters (common for all kinds of surface waters); and wetlands may receive seasonal overflow (most common on river floodplains).

Once accumulated in a channel or basin, surface waters have not completed their interaction with wetlands; surface waters in transit downhill often pass through wetlands. Some wetlands intercept surface waters only at high flows (e.g., wetlands on the floodplains of rivers), while others intercept all flow (e.g., a wetland created by a beaver dam on a small stream; figure 3-3). The main point is that precipitation moves downhill, and in doing so passes through wetlands either in transit to surface water or in transit along the drainage net after joining surface water.

Flow: The Soothing Effect of Wetlands

The three most important hydrologic functions of wetlands are (1) temporal spreading of flow, which leads to moderation of discharge volume and velocity; (2) spatial spreading of flow, which leads to moderation of velocity; and (3) maintenance of contact between groundwater and surface water, which leads to exchange.

The natural drainage network does not convey water efficiently from one point to another. Precipitation creeps through soil or groundwater often for weeks or even years before reaching daylight downhill as surface water.[5] Transit over the surface also is slow. Water in a channel often flows twice the distance of a straight line between two points, and may in transit enter the interstices of a stream bottom where the flow is much slower than in the channel. At the interface of upland and channel, or even lying across the channel, are lakes and wetlands, which impound water for varying times depending on their storage capacity.

No one seems to have estimated carefully the amount of water detention that is caused by wetlands, but a rough estimate suggests that surface water en route to the sea would on average spend one to two weeks within a wetland.[6] Thus, wetlands are an important part of the natural detention system that shaves peaks from the hydrograph (figure 3-4).[7]

A complementary hydrologic function of wetlands is to release water to streams and rivers after the highest discharges have passed. Thus, the storage capacity of wetlands not only suppresses the peaks of the hydrograph but also raises the valleys (figure 3-4).

The value of water storage by wetlands is to reduce flooding in wet

5. Soil water remains in the soil on average for approximately 280 days (Dooge 1984).

6. The contiguous area of the United States has an annual runoff of approximately 1.7×10^{12} m³/yr. The area of wetland in the contiguous states is about 5×10^{11} m². If a m² of wetland were to serve on average as passage for about 3 meters of water per year (a reasonable number), then the total area of wetland would be just sufficient to store and release the entire annual volume of runoff about once. The median residence time for water in wetlands, assuming annual average water depth of 30 cm (including water near the surface in saturated soil), would be one to two weeks. Behind the average, of course, is variation; some runoff is stored and released more than once, while other runoff may not reach wetlands at all, especially where wetlands have been removed.

7. For a case study, see Fusillo (1981).

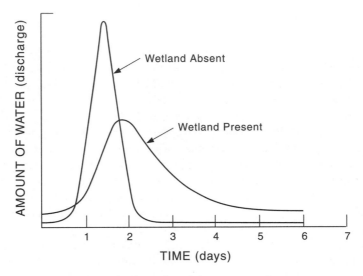

Figure 3-4. Diagrammatic illustration of the response of a stream to a storm event. Where wetlands have been removed, the storm generates more extreme discharge and quicker return to a lower baseline. The effect of wetlands is to spread the storm flow through time.

weather and to maintain the flow of streams and rivers during dry weather. Storage of water by wetlands also supports the growth of distinctive kinds of wetland organisms, such as cypress or muskrat. Any tangible or intangible value that these organisms have must be attributed in part to the storage of water by wetlands.

Water storage by wetlands is not beneficial from all perspectives in all situations. For example, water that is stored steadily loses volume to evapotranspiration. Where water is scarce, a wetland may release to the atmosphere water that is of considerable economic value. This is particularly true in the western United States, where most water is allocated to some purpose through a user of water.

A related, but not identical, function of wetlands is to reduce water velocity. Wetlands spread flowing water laterally when water is abundant. This reduces the erosive force of water.

Wetlands frequently are openings to groundwater; in fact, their lowest points may intersect the water table either constantly or seasonally when the water table is high. Through openings to groundwater, surface water enters (recharges) groundwater, or groundwater exits (discharges) to surface water. These exchanges stabilize the size of the groundwater reservoir. Recharge in particular can be important as the means by which groundwater removed for agricultural or other purposes is replaced with surface water.[8] Discharge of groundwater at the surface is also one means by which surface waters are sustained during dry weather.[9]

Reduction in the area of wetland within a given drainage causes the flow of surface water to be compressed in time, to show greater extremes of velocity and flow, and to interact less with groundwater. These changes usually are disadvantageous to humans.

Virtue among Waters

The finest of waters for human use are those that have low amounts of dissolved and suspended solids and negligible concentrations of substances that could be toxic to humans, aquatic life, or crops. We think of these waters as good. Other waters may be so saline that they kill plants and are unfit for consumption or support of freshwater fishes. Still others may contain so much suspended matter that they have a homely appearance or require special treatment before they can be used domestically. Some waters have hidden flaws: despite a fine appearance, they contain heavy metals or pesticides in such amounts that they are not safe to drink without special treatment and may be toxic to aquatic life. All of these waters are commonly considered bad.

Many bad waters lost their virtue because of things that were done to them. They received waste, were charged with sediment from naked soils, or became saline after being evaporated on croplands. Because people have compromised the virtue of so much water, it is easy to forget that some waters were bad, in the vernacular sense, from the beginning. The western United States, for example, has many naturally saline waters that are not useful for agriculture, domestic water supply, or support of most kinds of organisms.[10] In addition, weakly

8. Water users in the United States pump approximately 75-billion gallons of water per day from the ground. This is slightly less than one-quarter of all water that is used in the United States (Solley and Pierce 1992).

9. For a good example of relationships between wetlands and groundwaters, see Siegel (1988); for more general information, see Winter et al. (1998).

10. Invertebrates reaching significant abundance in the most saline arm of the Great Salt Lake include only the brine shrimp (*Artemia*) and the larva of the brine fly (*Ephydra*) except when wet years lead to dilution of salt, at which times the community becomes more diverse (Wurtsbaugh 1992).

vegetated soils in the arid parts of the nation produce runoff that is naturally laden with suspended solids.[11] Thus, while many waters have become bad through no fault of their own, others have been bad all along.

The amount and composition of dissolved and suspended substances in water are referred to collectively in the fields of engineering and environmental regulation as *water quality*. This traditional terminology is now a bit awkward because it suggests that waters should be evaluated in terms of their usefulness (i.e., their quality relative to consumptive uses of water). The chemical protection of water, however, is based on change in water quality rather than absolute water quality. While a water provider for cities may view the Great Salt Lake as bad water, which it surely is from the viewpoint of domestic consumption or agricultural use, we may view it from an ecological perspective as a reflection of the splendid variation of nature and perfectly suited to support the specialized organisms that live in it. As long as this water is not perturbed in its quality by human influence, we should regard it as ecologically good, even though it is bad from the viewpoint of consumptive use. In fact, the goal of the Clean Water Act is to reduce to negligible proportions the societal impairment of water quality.[12] Water quality then would vary through natural causes, without additional variation superimposed by humans.

The intent of the Clean Water Act, as expressed by Congress, is to "maintain the physical, chemical, and biological integrity of the Nation's waters."[13] Integrity is an odd concept to apply to water. At first it seems as though congressional staffers who crafted this phrase could have had an investment banking regulation on their minds. More careful thought shows, however, that the drafters may have been quite clever in adding a new word to the water lexicon. A new word allows latitude for interpretation without historical encumbrance. In fact, interpretation is still in progress (Angermeier and Karr 1994).

Even though the Clean Water Act mentions physical, chemical, and biological integrity, the nickname of the act (Clean Water Act) betrays its highest priority (clean water). It is significant that this act was not

11. The names of rivers in these regions (e.g., Dirty Devil and Colorado) indicate the indigenous nature of their high amounts of suspended solids.

12. Section 101 of the 1972 Federal Water Pollution Control Act states bluntly that the national goal is to eliminate discharge of pollutants into navigable waters.

13. This phrase appears in Section 101 of the 1972 amendments to the Federal Water Pollution Control Act.

called the Native Flow Preservation Act or the Aquatic Plant and Animal Act. The emphasis on clean water neatly combines many, but not all, of society's interests in domestic and commercial use, aesthetic appreciation, and natural productivity of water. When interpreted in a simpleminded way, as all laws must be at first, the highest priority of the Clean Water Act is water that is unimpaired by pollution. Hidden in the commitment to physical, chemical, and biological integrity, however, is a problem only partially acknowledged by present implementation of the Clean Water Act: absence of pollution guarantees neither physical nor biological integrity. A stream, or even a great river, may be reduced to a trickle by water diversions or may be physically impaired by channelization or disturbance. Thus, even the complete elimination of water pollution would not be sufficient to ensure the physical, chemical, and biological integrity of the nation's waters.

Wetlands are protected physically (i.e., they typically cannot be drained or filled without a permit) and thus are one of only a few examples for which the physical implications of the Clean Water Act are manifest. Ironically, even physical protection of wetlands is sustained primarily by the connection between physical properties of wetlands and maintenance of water quality rather than by the value of physical integrity for other purposes, such as maintenance of aquatic life.

Water quality is judged for purposes of the Clean Water Act through the "beneficial uses" of water, which typically include domestic and commercial supply, support of aquatic life, recreation, and agricultural supply. Prescriptions for improvement in water quality, the improvements themselves, and the requirements for evidence of improvement have been defined primarily in terms of kinds and amounts of dissolved and suspended substances in water (i.e., water quality).[14]

Because water quality is of prime importance to the Clean Water Act and because wetlands are protected by the Clean Water Act, it follows that wetlands must be important in the protection of water quality. This is commonly asserted but seldom explained.

14. For example, the water quality standards of Colorado, which reflect requirements of the Clean Water Act as administered by Colorado under review from the USEPA Region VIII, specify as chronic (30-day average) standards for arsenic the following: aquatic life use, 150 µg/l; agricultural use, 100 µg/l; and drinking water supply use, 50 µg/l. Any discharge of wastewater is restricted by a permit that requires compliance with these concentration limits in the receiving water. (Colorado Department of Public Health and Environment Water Quality Control Commission. Basic Standards and Methodologies for Surface Water 5 CCR 1002-31).

Quality: The Uplifting Effect of Wetlands

The effect of wetlands on water quality can be traced to a single important principle: detention of water within wetlands promotes physical and biological processes that improve water quality over time. Detention affects inorganic solids, organic matter, and dissolved inorganic substances.

Detention of water, with the attendant low water velocities that are characteristic of wetlands, allows the sedimentation of fine inorganic material that cannot settle within the channel of a swiftly moving stream or even a slowly moving river. The clarification of surface water by wetlands in this way is quite remarkable[15] and constitutes a major improvement of water quality that would be of great significance even in the absence of other processes.

The effect of wetlands on the organic matter in water is a bit more complicated. Wetlands receive, trap, produce, consume, and export organic matter. The balance of these five functions determines the effect of the wetland on concentrations of organic matter in water that exits the wetland (figure 3-5).

Ecosystems in general moderate extremes in the organic matter content of water that passes through them; wetlands conform to this principle. If a wetland receives water laden with a large amount of organic matter, it typically will reduce the amount of organic matter by providing the time and surface area for microbial decomposers to convert organic matter to carbon dioxide and water. If a wetland receives water that is extraordinarily free of organic matter, as may be the case for wetlands that are sustained by precipitation or groundwater, it typically will add organic matter that originates from the growth of higher plants and algae under wetland conditions. Thus, wetlands have a homeostatic effect on the organic matter content of water.

Because of their ability to stabilize the amount of organic matter in water, wetlands often compensate for human tendencies to cause the release of organic matter to water. A wetland may be regarded as a composting unit with reserve capacity to break down organic matter. This reserve capacity can be tapped in moderation without damage to the wetland. Thus, wetlands can absorb and process organic matter

15. Johnston (1991) showed that the mean amount of sediment trapped per year at 36 wetland sites studied by various methods was 1680 g/m^2 (15,000 pounds per acre). See also Brinson (1995).

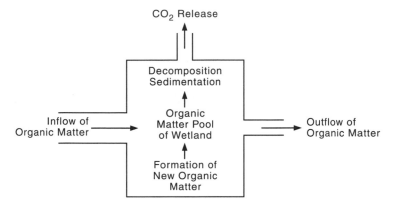

Figure 3-5. Pipe diagram of the processing of organic matter by a wetland that receives large amounts of organic matter from anthropogenic sources. Organic matter enters in large quantities and is joined by a small amount of organic matter that is formed by photosynthesis in the wetland. Part of the organic matter is converted to CO_2 and water by decomposition; the CO_2 is then lost to the atmosphere and in this way disappears from the wetland. For this reason, the output of organic matter is considerably less than the input.

that is liberated by a variety of human activities, such as rearing of livestock, dispersed drainage from fields or residences, or wastewater disposal, and in doing so maintain water quality at no cost beyond that of preserving the wetland.

Wetlands also modify the concentrations of substances that are in high biological demand (i.e., nutrients). Two of the most important nutrients are phosphorus and nitrogen.[16] Wetlands have a moderating influence on nitrogen, phosphorus, and other nutrients, just as they do on organic matter. Because water moves slowly through wetlands, there is much opportunity for physical and biological processes to affect concentrations of nutrients. Most often, the amounts of nitrogen and phosphorus in water are reduced by wetlands. This happens in several ways. Substantial proportions of nitrogen and phosphorus are attached to or carried by particles, which are removed by gravity or by filtration

16. A nutrient is any substance that is required for the growth of an organism. A subset of nutrients, called limiting nutrients, are those for which demand is likely to exceed supply. Limiting nutrients have the potential to stop growth when they become depleted. The two most common limiting nutrients in terrestrial and aquatic systems are nitrogen and phosphorus.

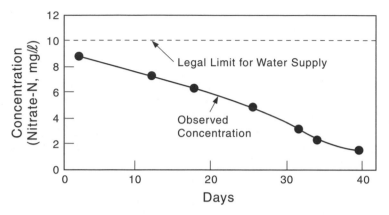

Figure 3-6. An example of the effectiveness of wetlands in removing nitrate. The solid line shows decline in nitrate concentration over wetland sediments from a high near the legal limit for water supply to a concentration that more nearly represents background conditions. This study was done on the floodplain of the Cache River, Arkansas, a tributary of the Mississippi (figure modified from Delaume et al. 1996).

through plants or soils in wetlands. The dissolved fractions of nitrogen and phosphorus also are removed because they are in strong demand by plants. Nitrogen is removed by yet another mechanism as well. The soils of wetlands are often anaerobic. Under anoxic conditions, microbes convert nitrate to nitrogen gas. This process (denitrification) leads to loss of nitrogen gas to the atmosphere (figure 3-6).

Waters that have high concentrations of nutrients are typically considered to be of low quality; they are designated *eutrophic*. Particularly when allowed to stand in a lake or pond, these waters show a constellation of traits that are the lacustrine equivalent of obesity. Eutrophic waters grow algae in nuisance proportions, have low transparency, suffer depletion of oxygen near the sediments, have high pH, fail to support some species of game fish, and produce unpleasant taste and odor. While the eutrophic condition is natural in some places, anthropogenic (cultural)[17] eutrophication of waters is widespread because human activities liberate phosphorus and nitrogen in large quan-

17. Eutrophic means richly nourished, i.e., receiving such abundant supplies of limiting nutrients that plant growth is nutritionally stimulated. Much eutrophication is induced by human activity; eutrophication of human origin has been called *cultural* eutrophication (Valentyne 1974).

tities through the disposal of wastes, use of fertilizers, combustion of fossil fuels, and disruption of the natural nutrient retention capacity of terrestrial ecosystems.

Wetlands are a defense against cultural eutrophication. They perform this function naturally by removing part of the nutrient load from the water source, thus reducing the mean concentrations of nutrients downstream.[18] They have the ability to absorb nutrient loads that exceed the natural or background level, thus compensating for human tendencies to liberate nutrients through virtually every aspect of life in an industrial society.

Organisms in the Wetland: No Place like Home

The Clean Water Act mentioned biological integrity offhandedly, as if this were a comfortable old notion among the biological cognoscenti. Biologists, meanwhile, could only search the act in vain for a definition; they had had no more experience with biological integrity than with biological depravity.

Fortunately, some useful definitions have been proposed, although the literature on the subject still gives the impression of work just beginning. Angermeier and Karr (1994) have captured the concept nicely as follows: "integrity refers to conditions under little or no influence from human actions." One might ask why integrity needs to be defined so restrictively. The answer lies in the connections between wetland organisms and their environment.

Even a novice Darwinist knows that organisms are adapted to a particular range of physical and chemical conditions and to specific kinds of competition and predation. A major change in any of these conditions, therefore, may suppress some kinds of organisms and some kinds of biological functions. Water quality is a case in point: changes in the quality of water often affect the composition of aquatic communities (e.g., invertebrates, aquatic plants, algae, and fish). Changes in aquatic communities may in turn affect other communities (waterfowl and mammals). Thus, the maintenance of water quality by wetlands is important to the integrity of all waters that pass through wetlands. Such waters are even now, after much loss of wet-

18. There are limits to the amount of cleansing that a wetland can do, especially if it is valued for support of aquatic life (Helfield and Diamond 1997).

lands, a high proportion of the total surface waters of the United States.

Biological functions of wetlands are numerous and diverse. They can be organized in a hierarchy that begins at the ecosystem level and extends downward in specificity to functions of individual species of plants, animals, and microbes.

At the ecosystem level, each wetland has characteristic biodiversity and characteristic rates of organic matter production (through photosynthesis), decomposition (degradation of organic matter), production of primary consumers and carnivores, and biological processing of nutrients. Within these ecosystem-level functions, we may seek more specificity by looking at groups of organisms or clusters of functions and ultimately at individual kinds of organisms and specific kinds of functions.

The biotic functions of wetlands have an assortment of values. As already mentioned, biotic functions that influence the transmission of nutrients across wetlands can have considerable value because they contribute to the maintenance of water quality by wetlands. Potential values, however, extend well beyond this. For example, wetlands produce fish, mammals, and birds that have value for commerce, sport, and aesthetic enjoyment. The values associated with commerce and sport are easy to understand but often are difficult to quantify. Aesthetic value, which is even more difficult to evaluate, is large and appears to be increasing.

Reckoning the Value of Biodiversity

As loss of wetlands progresses, species that are directly dependent on wetlands decline in abundance (Figure 3-7), and some even become extinct. People who are interested in these species, or in the diversity of life generally, are disappointed with these trends. It is not always clear, however, why loss or rarefaction of species should be of concern to the dedicated utilitarian.

Three kinds of arguments are used in explaining the value of biodiversity to those who may not value it for its own sake or on the mere assurances of others: (1) the stability argument, (2) the genetic reserves argument, and (3) biophilia.

Stability. The stability argument is easiest to explain by analogy. We could compare the diversity of species in an ecosystem to the di-

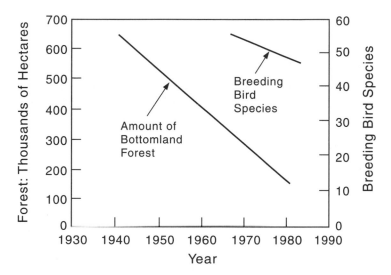

Figure 3-7. Reduction in the amount of bottomland hardwood forest in the delta area of the Mississippi River (Tensas River, Louisiana), and coincident loss in number of species of breeding birds along a road transect through the area (modified from Gosselink et al. 1990 and Burdick et al. 1989).

versity of players on a baseball team. The team has a premeditated diversity of morphotypes that perform different functions. Some have the ability to deliver a long ball with uncanny accuracy, while others are as quick as rabbits over short distances, and so forth. The team as a whole can, because of its diversity, defend itself against virtually any kind of challenge.

A very different sort of baseball team would be recruited under a peculiar set of rules allowing only the selection of catchers. The catcher's position would be filled perfectly, of course, and the team might do wonderfully at snagging low line drives. This team would not be competitive, however, because its excessive specialization could be exploited in a dozen ways by any more diverse team. In competition, a team of catchers would be unable to survive unless provided with endless subsidies by an eccentric billionaire who fancied catchers.

The stability of natural communities is also related to diversity, which may be defined loosely as variety in the kinds of organisms.[19]

19. Biodiversity encompasses more than variety of species; it also includes genetic variety within species and biotic variety above the level of species.

The relationship between stability and diversity is not yet well understood, but there is little doubt that major simplification of ecosystems through the interfering human hand does destabilize living communities (Tilman 1999). An example is a cornfield or rice paddy. While these monocultures are fabulously productive because they grow in absence of competition, they are inherently unstable; they require continuous addition of effort and resources in the form of pesticides, herbicides, and mechanical cultivation, without which they would be invaded by competitors and pests. Constant subsidy to maintain the stability of the monoculture may be worthwhile in agriculture because the subsidy can be exceeded by the value of the crop. Nevertheless, the instability of monoculture is quite clear.[20]

Monoculture is an extreme form of low diversity. Communities that are only slightly more diverse than a monoculture may be quite stable, even though stripped of most of their native species. The jury still sits on this issue, but it is clear that polyculture, as is practiced by a number of societies that mix their crop species as insurance against famine, is more stable than monoculture, although its maximum productivity is often lower.[21] Even a champion of biotic diversity would be forced to admit that the living world might be quite stable and productive with many fewer species. The terminus for human manipulation of the living world could be a sort of gigantic Japanese garden that requires regular maintenance but in return produces food and a stylized kind of beauty. Thus, the stability argument does not make a completely satisfactory case for maintenance of biodiversity, either in wetlands or elsewhere.

Most of the five-million or more species[22] on earth are commercially and culturally anonymous. Human attention focuses on a few

20. For example, intensive production of corn requires 7-million kilocalories per hectare of energy (as fuel, chemicals, drying, etc.) to sustain a yield of 24-million kilocalories per hectare in the form of grain (Pimentel 1984).

21. The agricultural systems of the United States and northern Europe are extensively based on monocultures sustained by high investments but even these systems typically involve the principle of crop rotation, which is a form of polyculture achieved by variation across years. Rotation is specifically designed to reduce the costs of monoculture, some of which result from accumulation of diseases and pests that are well adapted to attack a large stand of a particular crop species (Altseri 1995).

22. Approximately two-million species have been described (i.e., they are known to science), but the total number of species is much larger. Until the early 1980s, the total number of species had been estimated as about five million, but it now seems plausible that the number could be five or more times higher (Wilson 1988, Rosenzweig 1995).

hundred species of star quality that provide food, fiber, and medicine.[23] For hard-core utilitarians, the living world might be perfectly adequate if it supported only these species. These species, however, cannot live alone. Star plants require pollination as well as animals that can disperse their seeds and specialized fungi that help them draw nutrients from the soil into their roots. Star animals require plants or other animals for food and vegetation for shelter. The utilitarian might suggest that we make a list of the supporting cast and maintain it as well as the species of primary interest. Each member of the supporting cast, of course, has its own supporting cast; thus, the utilitarian becomes an advocate for diversity. Some portion of the supporting cast (presumably the extras) can probably be lopped off, but it is exceedingly difficult to make an accurate cut—and inaccuracy could spoil the whole objective, even if the goal were as narrow as the maintenance of species used commercially by society.[24] A few species can be maintained by intensive management in the form of animal husbandry or agriculture, but the rest must live with much smaller subsidies.

Genetic Reserves. Even though many commercially important species cannot be maintained in the absence of a biotic web of other species, it still may be possible to eliminate selectively some kinds of communities that lack commercially important species, or to bring the commercially important species increasingly into culture until they could be weaned from their mutualistic dependencies on species other than humans. This would allow much simplification of nature. The immediate losses might be tolerable, but the capital losses could be staggering. Only a fool would argue that humans have already thought of every way in which they might use other organisms (Myers 1983). The continuing flow of new pharmaceuticals and biochemicals from tropical plants shows that there is vast unexploited potential benefit among the millions of species that are yet unknown and unnamed, as well as in the one million or so that are known but still unfamiliar.

23. About 85% of human food comes directly or indirectly from 20 kinds of plants; just 3 kinds of plants provide two-thirds of the total (corn, wheat, and rice; Raven 1988).

24. The Biosphere 2 experiment in Arizona showed how difficult it would be to engineer a simplified life support system, even with unlimited resources. Biosphere 2 involved the creation of a 3.15-acre enclosure at a cost of approximately 200-million dollars. After two years, both oxygen and carbon dioxide concentrations were severely deviant from external concentrations, and nitrous oxide reached levels that could lead to brain damage in humans; 19 of 25 vertebrates originally stocked in the enclosure became extinct, as did all pollinators (Cohen and Tilman 1996).

Now that genes can be plucked from one species and spliced into another, the concept of genetic reserves or genetic libraries makes a compelling case for preservation of even obscure species.

Biophilia. The final argument for maintenance of biodiversity is *biophilia*, by which is meant the innate emotional attachment that humans have to other kinds of life. Professor E. O. Wilson (1984) has helped us realize that much of our preservationist impulse may have come not through commercial concerns or even our desire to preserve genetic capital, but rather through the comfort and enjoyment that we draw from nature. We are free to put a value on biophilia and to tax ourselves to maintain it. It makes no less sense to pay for the ornaments of nature than for the aesthetic embellishments of architecture or for art itself. In fact, the aesthetic dimension of nature may be more deeply imbedded and more broadly accessible than any other kind of aesthetic pleasure.

Biodiversity, Functions, and Values. The three arguments for biodiversity can be translated easily into the jargon of functions and values for wetlands. Functions of wetlands include the support of groups of organisms, including species that are important presently for commerce and sport, as well as species that are not. The species that are not used directly by society support those that are, and also offer undeveloped potential for human use. Beyond the current and potential utility of wetland species is the aesthetic value of species that disappear when wetlands are lost.

Answering the Question, Finally

One may accept, unless one is hopelessly skeptical, that wetlands carry out numerous functions and that some of these functions have considerable value. Uplands, however, also perform numerous functions, some of which are of high value. Thus, a list of functions and values for wetlands does not in itself demonstrate why wetlands should be subject to unusual degrees of protection. A satisfying answer must involve other factors, including the extent and location of wetlands on the landscape and the capacity of wetlands to regenerate when altered.

Scarcity. For the United States as a whole, wetlands have, in a sense, always been scarce. Wetlands originally accounted for approximately 5% of the area of the contiguous United States but now have been reduced to about 2.5%.[25] To the extent that wetlands have distinctive functions and values, these are focused on a tiny proportion of the area of the United States. Because a scarce resource is more likely to be eliminated incidentally or inadvertently than a more abundant one, scarcity itself is a justification for special protection of wetlands.

The scarcity argument is actually more complex than mere acreages would indicate. Within the category of ecosystems that we call wetlands are distinctive classes or types. For example, swamps, marshes, and bogs differ from each other in ways that are obvious even to a casual observer (Mitsch and Gosselink 2000). Even within these large categories, however, there is a great deal of variation, particularly of a regional nature. A coastal marsh in North Carolina, for example, differs radically from a prairie marsh in North Dakota by virtue of its hydrology, substrate, and biota. On an even finer scale, we may find, within a region, substantial differences among wetlands of a given class. Wetlands that are inundated for long intervals, for example, may hold different species of organisms and have different hydrologic properties than wetlands inundated for only short intervals. Thus, the problems derived from scarcity of wetlands as a whole are exacerbated by variety among wetlands.

The significance of scarcity is probably as much instinctive or cultural as it is rational. Notwithstanding the constantly unfavorable press over its low rate of savings, American society probably contains more ants than grasshoppers, and ants abhor scarcity. If upland forest were to reach the same scarcity nationally as wetland, it too would probably be the subject of federal protection on grounds of scarcity alone.

The scarcity argument has some of the characteristics of supply and demand as applied to an nonrenewable resource such as petroleum. As petroleum reserves are depleted, petroleum becomes more valuable. The increase in value with depletion of quantity is progressive rather than steady: a given amount of depletion has much more significance when the resource is mostly gone than it does when the resource is unexploited. The same principle applies to wetlands. For example, wetlands of a given region may be important as a hydrologic

25. The contiguous 48 states have about 100-million acres of wetlands and the following millions of acres of land in other categories: forest, 395; range, 399; pasture, 125; crops, 382; water, 49; developed land, 92; and federal land composed of various natural and seminatural landscape types, for a total of 1891-million acres (USDA Natural Resource Conservation Service 1996).

buffer. Initial losses of wetland acreage may have subtle effects on flood flows, but continuing loss of wetland acreage will have an increasingly drastic effect on flood flows as the hydrologic buffer is removed. Thus, the value of remaining wetland acreage increases disproportionately as the total acreage of wetland decreases.

For people who live where dogs are eaten by alligators and everybody owns hip boots, the scarcity argument may be hard to accept. Congress attempted to deal with this issue directly in 1995 when the House of Representatives included in its draft amendments to the Clean Water Act a provision stating that no more than 20% of any U.S. county could be fully protected as wetland under the Clean Water Act. Given that much of the wetland area of the United States occurs in large blocks, such a rule would have released a large percentage of the total remaining wetland inventory of the United States from federal protection. Although the provision was removed, the underlying question remains: Is scarcity a local issue or a national one?

There is probably some merit to evaluating scarcity on a local rather than a national basis, but not enough to justify rules such as the one considered by the House in 1995. One must remember that large areas of wetland are located where there is a large amount of water. All functions that relate to the physical, chemical, and biological processing of water require high capacities in these regions. In other words, big wetlands serve big waters and should not be regarded as excessive from the viewpoint of hydrologic buffering or maintenance of water quality. In addition, large wetlands support quantities and types of organisms that would not be found in smaller or more fragmentary wetlands.

Location. Realtors are fond of saying that the three most important characteristics of housing are location, location, and location. We may grudgingly credit realtors with proper appreciation of a major principle in ecosystem science. Locations of wetlands are consistently strategic to the landscape as a whole because wetlands often connect land to water.

As explained earlier in this chapter, the locations of wetlands generally fall into three categories: (1) on or at the fringe of the drainage net; (2) on an overflow connection to the drainage net (e.g., a floodplain); and (3) isolated from the surface part of the drainage net but typically with a connection to groundwater. Each of these locations is a hydrologic junction that regulates the quantity and quality of water passing through it. At the fringes of the drainage net, wetlands are the gates through which seepage and overland flow pass from upland to streams, rivers, and lakes. Wetlands moderate both the flow and the quality of water and thus regulate the characteristics of water within the drainage net. Where water is flowing most rapidly, wetlands in the overflow zone provide reserve hydraulic and metabolic capacity that supplements the moderating influences of the fringe wetlands. Wetlands isolated from the drainage net have a similarly moderating influence on groundwater. Thus, the location of wetlands is strategically critical to the welfare of ecosystems beyond the wetland boundary.

Regeneration. Use of an upland forest for agriculture involves removal of trees followed by cultivation. In most parts of the United States, however, cessation of cultivation for even a few years leads to spontaneous reestablishment of forest. In fact, the reestablishment process, which is known as secondary succession (secondary because it occurs where a forest has existed before), is so important and widespread that it was a major theme in the early development of plant community ecology and terrestrial ecosystem science (Golley 1993). In the southeastern and midwestern United States, the landscape is rich in forests on lands that were once cultivated. These forests are distinguishable from the precolonial forests only in their deficiency of very old trees and by the presence or absence of a few species that are the unintended consequence of human commerce in exotic plants and their attendant pests and diseases. All in all, however, a regenerated forest in Illinois or North Carolina is distinguishable from the original only to a trained eye. It is robust, diverse in plant and animal life, and capable of self-maintenance. In other words, it is a healthy native ecosystem.

Grassland ecosystems of the United States are much like forest ecosystems in their ability to regenerate, although they have been more subject to invasion by exotic species than most kinds of forest (e.g., Baker 1989). Fields that lie fallow for only a few years spontaneously develop increasing biological diversity in the form of mixed grass species, a wide variety of insects, and small mammals. In this way, the grassland regenerates many of its original characteristics through secondary succession.

It is fortunate that many forests and grasslands of the United States have a high capacity for regeneration through secondary succession. Vast acreage in the United States has been cultivated and subsequently left fallow because initial judgments of profitability were wrong or because the economic basis for agriculture shifted in such a way as to change profitability to loss. Secondary succession has followed patiently in the footsteps of agricultural abandonment, sowing diverse plant communities and their dependent animal faunas over the top of abandoned monocultures. Without this aggressive regeneration of terrestrial ecosystems, many parts of America would be a moonscape of abandoned agricultural projects. The high regenerative capacity of terrestrial ecosystems in much of America is in part a happy accident caused by the tendency of temperate ecosystems to store large amounts of organic matter and nutrients that make up an ecosystem trust fund capable of subsidizing regeneration from bare soil and seed.[26]

The use of wetlands for agriculture or development is fundamentally different from the use of upland forests or grasslands; it requires drainage or a combination of drainage and filling. Because specific kinds of hydrologic conditions are essential to the existence of wetlands, drainage and filling often will preclude the spontaneous regeneration of wetlands. In some instances, wetland hydrology may reestablish itself spontaneously following abandonment of use, in which case regeneration is possible. In many cases, however, the establishment of drainage is accompanied by increasing human commitments to the availability of dry land (e.g., by installation of housing or roads), and drainage tends to be maintained even after its original purpose has become moot. Thus, the conversion of wetlands is much more likely to be permanent than the conversion of upland forests or grasslands.

The Logic behind Special Protection. Taken together, the scarcity of wetlands as a percent of the total landscape, the strategic location of wetlands between land and water, and the tendency for conversion to be irreversible except through active intervention that in many instances would be economically unfeasible, argue powerfully that wetlands require special protection if their characteristics and functions are of any significant value to society. The value of wetlands is hard to deny given clear evidence that they moderate the flow of water while improving its quality, and that they maintain distinctive complexes of organisms that cannot occur elsewhere.

26. At tropical latitudes, the soil trust fund is more often absent, and removal of the vegetation leads to irreversible biogeochemical changes that prevent regeneration of the original plant community (Schlesinger 1997).

4

WATER IS AS WATER DOES

The greatest subtlety of wetlands lies in their hydrology. This is unfortunate, given that the very name wetland invokes images of water on the land. In fact, those who have been impatient with the thicket of indicators by which federal agencies identify wetlands have at times demanded a return to basics in the form of simple inspection for water. One Washington insider even suggested that we might send out her Uncle Dennis to delineate wetlands.[1] If Uncle Dennis, who has a propensity to sit down, returns with a wet bottom, he has been in a wetland; otherwise he has not.

The truth is that Uncle Dennis could sometimes wet his shorts in an upland forest or grassland, yet at other times sit comfortably dry on a certified wetland. The hydrologic conditions that separate wetlands from uplands in fact involve four phenomena related to saturation:[2] proximity to the surface, time of year, duration, and frequency (figure 4-1).

saturation

1. Apparently, the Uncle Dennis concept was created by Lujana Welcher, Assistant Administrator for Water during the Bush administration. It struck a responsive chord in Washington at that time (M. N. Strand, personal communication).

2. A soil may hold a great deal of water without becoming saturated; saturation occurs only when the air voids between soil particles are completely filled with water. Furthermore, inundation of a dry soil does not always produce instant saturation because the soil may hold air voids for some time, even after sufficient water is present to produce saturation (Brady and Weil 1996).

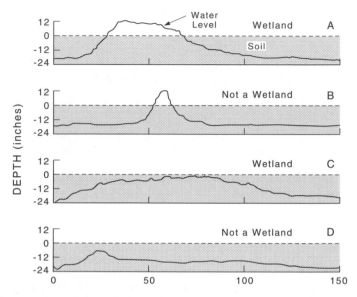

Figure 4-1. Illustration of some hydrologic conditions characteristic of the site in each case that relate to the formation of wetlands: (a) prolonged inundation during the growing season (wetland); (b) brief inundation (not a wetland); (c) prolonged saturation of the upper soil layer during the growing season (wetland); and (d) brief saturation of the upper soil layer within the growing season (not a wetland).

Proximity: A Matter of Inches

Proximity of saturation to the soil surface is one of the most widely misunderstood aspects of wetlands. In many wetlands, water stands on the surface for extended intervals or, in a few cases, continuously. The significance of standing water (inundation), however, is easily over-rated. Inundation is neither necessary nor sufficient for a wetland. Recurrent and prolonged saturation of soil near the surface is sufficient to create and maintain a wetland, even if water never actually stands on the surface. The exact distance within which saturation must approach the surface varies from one region or one wetland type to another but by rule of thumb is about one foot or, scientifically speaking,[3] 30 cm.

---

3. Use of English units here and elsewhere is not an endorsement of them, but rather a concession to persistence of English units in regulatory practice and legislation.

While plant roots often extend beyond one foot,[4] prolonged saturation of the upper foot blocks penetration of oxygen into the soil, and thus to roots, so completely that most plants cannot survive. Plants that tolerate or thrive in saturated soils are called *hydrophytes*; they are characteristic of wetlands. Thus, the critical depth for saturation is determined by the differential responses of hydrophytes and non-hydrophytes to prolonged immersion of roots in water.

The upper one foot of soil is also the zone of most intense microbial metabolism in soil. Except when the soil is very dry, soil microbes perform composting (decomposition) functions. The organic matter that supports composting comes mostly from plants, which may range from trees to ankle-high herbs and grasses. Microbial metabolism is most intense near the soil surface, where the organic feedstock originates, and fades away at greater distances from the surface.

Both plant roots and microbes use oxygen. Thus, the upper one foot of soil may lose its oxygen very quickly when the external supply of oxygen is blocked by water in the soil. As explained more fully in chapter 5, microbial metabolism continues even after the depletion of oxygen but in a different (anaerobic) mode. Microbial metabolism under anoxic conditions leads to changes in the color, composition, and structure of the soil. Thus, when soils are saturated with water, microbial activity affects the growing conditions for plants by using up oxygen (thus favoring hydrophytes) and also leads to the formation of a specific class of soils (hydric soils).

The one-foot rule, which is used in regulatory delineation through the 1987 Corps manual, has an embellishment that is both minor and significant, much in the manner of a pinhole leak in a radiator. The depth at which water stands in a hole in the ground is the depth of the water table. In substrates of coarse grain, such as sands, the water table marks the boundary of the zone of saturation (i.e., there is no water saturation above the water table). In soils that have fine pores, however, water is wicked upward by capillary action, in the way that a paper towel wicks water from a kitchen table. Thus, in soils of fine grain, the zone of saturation may extend above the water table.

The 1987 Corps manual gives zero as the critical depth for saturation of a wetland soil (i.e., water must reach the surface). In ap-

4. Maximum rooting depths for three major categories of plants are as follows: trees, 7.0 meters; shrubs, 5.1 meters; and herbaceous plants, 2.6 meters (Canadell et al. 1996).

plying this rule, however, the Army Corps, with support from the EPA, has assumed the water table threshold for wetlands to be 12 inches below the surface. Thus, the rule (surface saturation) appears to differ from the field test of the rule (water table within 12 inches of the surface). This difference has caused great consternation among members of the regulated community, some of whom believe that the surface saturation rule is being transmuted by jackbooted regulators into something completely different. There is an explanation, of course. The Corps has assumed that water will rise 12 inches from the top of the water table to the surface of the soil because of soil capillarity.

The assumption that water consistently rises through soil 12 inches from the top of the water table is a simplification. It is neither always correct nor is it always incorrect. Water may rise negligibly, a few inches, 12 inches, or possibly even more through capillarity. Thus, it is unrealistic to expect the water table to mark the upper limit of saturation and it is unreasonable to assume that water always rises 12 inches above the water table.

The NRC Wetlands Committee, which reviewed the issue of saturation (National Research Council 1995), suggested a modification of the present approach, as follows: (1) the critical zone for saturation extends one foot below the soil surface (as presently assumed); (2) the water table marks the boundary of saturation, provided evidence to the contrary is absent; and (3) when field evidence shows upward extension of the zone of saturation above the water table, water table measurements should be corrected accordingly. This approach dissolves the fiction about a uniform one-foot capillary zone, acknowledges the significance of saturation anywhere within the uppermost one foot of soil, and leaves open the possibility that saturation may extend upward from the water table. A change of this type at least would avoid the appearance of inconsistency between the rule for saturation and its interpretation in the field.

Time of Year: A Most Unusual Season

Organisms run on chemical reactions, the rates of which increase with temperature. As a rule of thumb (the temperature–metabolism rule, or Van't Hoff rule), metabolic rates double for a $10°C$ rise in temperature.[5] This rule breaks down at high temperatures (e.g., above 35°C), when proteins begin to become unstable, and near freezing, when the metabolic machinery of many organisms becomes so sluggish that it virtually stops. The temperature of an organism is determined by the temperature of the surrounding air, water, or soil, plus some direct solar heating of tissues exposed to sun. Exceptions, of course, include mammals and birds, which hold nearly fixed temperatures, but these account only for a tiny fraction of the Earth's total biotic metabolism.

Because metabolism depends on temperature, the annual amount of metabolic work that occurs inside an ecosystem is related to the number of days that the ecosystem spends at each temperature within its annual range. An ecosystem that is warm for many days thus can perform more total metabolic work than a perennially cold one. This explains why the coldest climates have the lowest potential to produce crops.[6]

The temperature–metabolism relationship is embedded in modern agricultural practice. In the United States, the influence of temperature is expressed in terms of growing season, which is the number of days that most crop species show agriculturally meaningful growth.[7] The length of the growing season determines whether a given crop species will be able to mature and, for the warmest climates, whether multiple cropping will be possible.

The concept of growing season has been used by federal regulatory agencies in defining the hydrologic conditions that lead to formation of wetlands. Specifically, saturation of soil with water is considered relevant to the formation of wetlands only if it occurs during the growing season. Saturation of soil at or near the surface causes the formation of wetlands because saturation selectively stimulates or suppresses plants, microbes, and other kinds of organisms. When the soil is so cold that biological activity slows greatly or stops, organisms

5. The rise in metabolism may actually be greater or less than the rule of thumb would suggest, but a substantial increase in metabolism is likely to accompany a substantial increase in temperature (e.g., Prosser 1991).

6. Mean annual production of dry matter in natural plant communities is about 500 g/m² where the annual mean temperature is 0°C and 2500 g/m² where the annual mean temperature is 25°C, if water is available (Leith 1973).

7. Agricultural growing season is typically expressed in terms of the length of time between killing frosts. For most crops, the growing season measured in this way must exceed 120 days (Martin et al. 1976).

may not respond, either positively or negatively, to the presence of water. Thus, saturation of the upper soil layers outside the growing season, even for prolonged intervals, is not by itself sufficient to maintain a wetland. Therefore, it seems reasonable to disregard saturation of the upper soil layers at times of biological inactivity for purposes of identifying the hydrologic conditions that sustain wetlands.

Despite its seemingly rational foundation, the growing season concept has proven to be a needle's eye for hydrologic definition of wetlands; it suggests that wetlands cannot occur in some cold regions where they do in fact occur in abundance. A closer look reveals flaws in the concept.

The transition from nongrowing season to growing season is assumed to occur when the mean soil temperature passes a threshold that has been called *biologic zero*, which is assumed to be 5°C or, alternatively, the interval extending three weeks before to three weeks after a killing frost.[8] Days of saturation below the biologic zero or, alternatively, outside the boundaries set by the three-week rule, are discounted. In other words, all ecosystems are assumed to have the same temperature threshold for activity.

The growing season concept works for agriculture because crop species, which have been chosen partly for their productivity, require an extended period of warmth. The situation is very different for wetland plants, which are more diverse metabolically than crops. In cold regions such as Alaska, wetland plants include species that are adapted to complete their life cycle where the temperature never rises above the biologic zero that is now used for wetlands.[9]

Growing season has proven to be a surprisingly troublesome concept, but remains indispensable in the identification of wetlands. The problem of growing season has two parts: (1) separation of growing season from the rest of the year and (2) recognition of changes that occur within the growing season.

---

8. The 1987 Corps manual uses both the 5°C threshold for biologic zero and the three-week rule for identification of growing season. The 5°C rule is applied at a soil depth of 20 inches. The three-week rule differs from standard agricultural definitions of growing season because wetland plants may be active after crops (which are mostly temperature-sensitive annuals) are inactive or dead.

9. The most striking exceptions probably occur in permafrost wetlands, which have mean annual soil temperatures below 0°C and in some locations show an official growing season of zero days, even though they support an abundance of wetland vegetation (Bedford et al. 1992).

The first issue arises because organisms that live in cold climates have evolved enzymes that function efficiently at low temperatures. For this reason, the assumption that all ecosystems will show negligible metabolic activity at the same temperature cannot be correct. In particular, cold climates will support plants, microbes, and other organisms that continue to function significantly (although at reduced rates) at temperatures near or below 0°C. Because these ecosystems spend long periods of time at temperatures near zero and because they may be quite active at such low temperatures, their prolonged saturation with water can be sufficient to establish wetland conditions even when temperatures are near 0°C. In other words, growing seasons must be defined regionally and with reference to natural communities rather than crops. There is no single biologic zero.

The second issue related to growing season derives from change in soil temperature during the growing season. The effect of water saturation on a soil depends on temperature and thus varies during the growing season, as explained below.

Duration: Relativity Theory for Wetlands

Because rates of metabolism increase as temperature increases, the effect of saturation accelerates as temperature increases. For example, 10 days of saturation in summer, when the soil is warmest, are typically sufficient to cause depletion of oxygen in soil and thus produce conditions that lead to the establishment of a wetland (figure 4-2). The same duration of inundation in the early spring, when the soil is cooler, may not lead to the formation of a wetland because metabolic processes would be proceeding more slowly and oxygen would be depleted more slowly. Thus, the challenge is to relate the threshold for soil saturation to the temperature of the soil. This would be possible by use of duration tables showing change in critical duration with change in temperature or by other means combining time and temperature.[10]

The growing season for wetland plants and soil microorganisms

10. The need to combine time with temperature in analyzing metabolic processes has long been appreciated in biology and agriculture. Examples include the use of degree-days or total accumulated heat to predict growth of both plants and animals (e.g., Gates 1993).

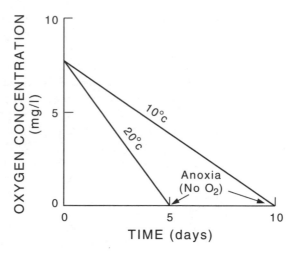

Figure 4-2. Illustration of the effect of temperature on depletion of oxygen in soil. Oxygen often is depleted from cool soil over a period of approximately 10 days of saturation. An increase in temperature of 10°C will reduce the time of depletion by about half, provided that the soil contains substantial organic matter.

must be defined relative to climate because organisms growing in cold climates may be metabolically active at low temperatures. When a growing season has been defined in this way, the significance of changes in soil temperature during the growing season must also be acknowledged because biological processes occur more swiftly after soil has warmed fully than in early spring while it is still cool. Thus growing seasons ultimately must be defined by region, and water saturation during the growing season must be interpreted on the basis of soil temperatures at the time of soil saturation.

Frequency: The Eternal Return

An unusual flood could inundate ground that is seldom flooded. If inundation were prolonged during warm weather, it could be sufficient to kill or harm some or all of the upland plant species and to cause anoxia in the soil. Still, the inundated zone would not be a wetland because recession of the flood waters would be followed by reestablishment of upland plant species, which in time would return as the dom-

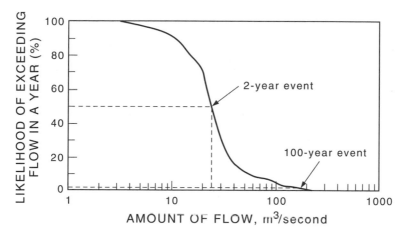

Figure 4-3. Recurrence graph for the South Platte River below Denver, Colorado, showing the 100-year flood (a very unlikely event) and the two-year flood (an event of sufficient frequency to maintain a wetland).

inant vegetation. Given that anoxia in the soil had been unusual, there would be no fundamental change in the soil, which would retain its upland character. Mammals and invertebrates requiring unsaturated soil could be eliminated or forced out by the flood, but would return. Thus, an unusual hydrologic event does not create a wetland.

Hydrologists describe the frequency of hydrologic events in terms of their recurrence interval (Hornberger et al. 1998; figure 4-3). Recurrence in this context is a probabilistic concept, i.e., it applies to averages and not to individual events. The irregularities of probabilistic events can be confusing to flood victims. Those who have experienced losses to 100-year floods twice or more in a lifetime may question the recurrence interval, but it is true that a period of 200 years may produce no 100-year flood, whereas another period of 200 years could produce four 100-year floods. The concept of recurrence is essential to an understanding not only of major floods but also of the more frequent high-water events that maintain wetlands.

Saturation at or near the soil surface for an extended interval every year during the growing season will sustain a wetland. When saturation occurs only at intervals of decades or centuries, a wetland cannot be sustained. Between these extremes is a frequency threshold that divides wetlands from uplands. The frequency just sufficient to maintain a predominance of organisms that occur in wetlands and cause the development of wetland soils (hydric soils) is the threshold frequency for wetlands. As a rule of thumb for federal regulation, the threshold frequency is one out of two years, which corresponds to a recurrence interval of two years, i.e., an area that is inundated or saturated near the surface for an extended interval during the growing season for at least one out of two years over the long term is likely to be a wetland.

Dismissing Uncle Dennis

It should be clear at this point why Uncle Dennis will find little work as a hydrologic consultant. His sensor, though sensitive, tells us nothing about the subsurface, which is hydrologically critical in distinguishing wetlands from uplands. In addition, his data collection is sporadic, while the hydrologic distinctions between the prolonged saturation characteristic of a wetland and the brief saturation that might occur on an upland can be judged only from routine and frequent acquisition of data. Finally, Uncle Dennis's observations in a given year might or might not be characteristic of a long run of years, as would be critical for the evaluation of wetland hydrology.

For the same reason that we have dismissed Uncle Dennis, we would also dismiss any measurement, however technical or scientific it may appear to be, that gives us information for only one year, or only at irregular or widely spaced intervals within a given year, or that deals with only the surface. For example, we might note that a given area had flooded the previous spring or we might find drift lines or blackened leaves where the water had stood and then receded.[11] In such a case we would know that the surface had flooded, and we could assume that the underlying one foot had been saturated, but we could not know whether this flooding was characteristic or unusual.

With Uncle Dennis's deficiencies in mind, one could devise, with

---

11. These are not hypothetical examples; they have figured importantly in the evaluation of hydrology for purposes of wetland delineation by the Army Corps of Engineers. A hard-nosed hydrologist would point out that the presence of blackened leaves or drift lines indicates the recent presence of water, but not the repeated presence of water. These so-called *surface indicators of hydrology* are, however, not useless. For example, if they are found on a site in a year of average moisture, it would be reasonable to deduce that the surface of the site typically is saturated in years of average or above-average wetness.

the assistance of a hydrologist, a means of collecting hydrologic information that would nicely define the depth, timing, duration, and frequency of soil saturation as necessary to separate wetlands from uplands. There are three direct approaches to the assessment of hydrology: (1) extended measurement of water table depth, (2) limited measurement of water table depth combined with computer modeling, and (3) photography. The first of these is the most rigorous and satisfactory approach; the second is sometimes satisfactory but more complex and not always feasible; and the third is better than nothing but has numerous flaws.

Extended measurement of water table depth can be achieved by the installation of shallow wells that are equipped with water level recorders. Depth to the water table thus can be recorded continuously, and an upward correction can be added if the soils in question cause, by capillary action, saturation to reach significantly above the water table.

The problem with a direct assault on the hydrology question is that the collection of data must occur at high frequency and at multiple locations for several years. Thus, this approach is essentially useless for most regulatory delineations.

A second way to assess hydrology is to collect a small amount of data (e.g., weekly data for three shallow wells over one year) and use computer modeling to estimate what this data set would look like if it were collected over many years. Estimations of this sort are sometimes feasible because water table depth is related to weather (especially precipitation), and weather is measured at many locations over the long term. Thus, a computer connection between a long-term weather record and a short-term hydrologic record can produce an estimate of a long-term hydrologic record (National Research Council 1995). There are problems, of course. The hydrology of some locations is so complicated that the connection between water table and hydrology cannot be predicted very well. In other instances, hydrology has been changed or is changing because of drainage or diversion projects that complicate modeling.

Short-term data collection coupled to modeling is more practical than long-term data collection but still is impractical for everyday use. Although it does not require many years of data, it does require measurements spanning at least several months, which can seem like a long time to someone who is losing money by the day. In addition, hydrologic modeling can be outrageously wrong if used incorrectly. In the long run, this approach will be useful in selected circumstances but not for most of the delineations that must occur every day as long as the Clean Water Act continues to require permitting.

The third way to collect hydrologic information directly is by photography, which is the most practical of the three direct methods. In principle, one could photograph a landscape repeatedly and, over a long period of time, accumulate information on the extent and frequency of inundation. Satellite photographs could be used for this purpose but would provide information that is too coarsely resolved for most delineation purposes. Airplanes, however, can collect information of proper spatial resolution.

The virtues of photography for assessing surface hydrology have been known for a very long time but aerial photography was not brought into the routine identification of wetlands until the USDA's Natural Resource Conservation Service (formerly the SCS) was required, through the Food Security Act of 1985, to map wetlands on agricultural lands. This requirement was so overwhelming in scale that it could be met quickly in only one way, which was photography. It turned out that the NRCS was taking annual photographs already for the assessment of crop acreages. This program had been in place for many years as a means of validating the crop acreages that are allowed to landowners under agricultural subsidy programs. The NRCS hit upon the clever idea of estimating the frequency and extent of inundation for a given area by looking at the crop assessment photographs of that area over multiple years. This has become the primary means by which the NRCS maps wetlands on agricultural lands. It is not used by the Army Corps for nonagricultural lands.[12]

Photography on an annual cycle, as employed by the NRCS, to some extent resolves the vexing problem of estimating recurrence because it provides direct information on recurrence of saturation. Ideally, however, photography would be repeated several times each year during the season of inundation in order to demonstrate the duration of inundation. The NRCS typically uses only one set of photographs for

12. There are at least two reasons. First, the NRCS was charged with mapping all wetlands on agricultural lands, whereas the Army Corps has had the leisure (in a relative sense) of mapping only wetlands affected by permits or, more often, of checking maps made by consultants paid by landowners requesting permits. In addition, delineations dealt with by the Army Corps often involve areas for which there is extensive tree canopy or other vegetative cover that would be more troublesome for remote sensing than croplands.

each year and then attempts to estimate duration from signals such as yellowing of crops, but this is an imperfect substitute for multiple photographs within a given year. Thus, the present NRCS photographic scheme helps resolve the recurrence problem but not the duration problem. An approach that would show duration directly by photographs would be much more expensive.

One other shortcoming of the photographic approach is that it only gives information about the surface. Because the relevant hydrologic conditions for wetlands extend downward 12 inches from the surface, photography is a crude way of assessing hydrology for the purpose of identifying wetlands. Experienced photointerpreters often can detect effects of subsurface saturation through the appearance of vegetation but are handicapped in doing so by the availability of only one photograph per site per year.

Divining Rods for Wetlands

Nature records its own history, as we well know from fossils, erosional landforms, and layered sediments. Given that direct hydrologic evidence for the presence of a wetland is seldom available, it is reasonable to search for information that will reveal the hydrologic past of a given site. Two kinds of natural records have been interpreted for this purpose: soil and vegetation.

In anticipation of chapter 5, which will show what soils have to tell about wetlands, and chapter 6, which does the same for vegetation, it suffices to say here that the prolonged presence of water during the growing season at or near the terrestrial surface changes soils and vegetation in ways that can be observed during seasons or years when the soil is not saturated. Thus, the special soils (hydric soils) that are associated with repeated and prolonged saturation and the special vegetation (hydrophytic vegetation) that is associated with the same conditions can be taken as evidence for the prior existence of wetland hydrologic conditions. In the language of delineation, hydric soils and hydrophytic vegetation are *indicators* of the wetland hydrologic condition.

Some critics of the present system of wetland regulation have suggested or insisted that the hydrologic characteristics of a site be assessed directly rather than indirectly through an analysis of soil or vegetation. In fact, this was one of the major inspirations behind the

1991 proposed revisions to the 1987 Corps manual. Proponents of change claimed that the three kinds of evidence (water, soil, and vegetation) for the presence of wetlands must be assessed independently; they argued that three independent kinds of evidence provide a more secure diagnosis than two kinds of evidence.

There are two reasons to oppose a general requirement for direct hydrologic information in support of wetland delineation: (1) it is impossible and (2) it is unnecessary. Those who claim that direct hydrologic information would prevent some errors of inclusion in the analysis of wetlands may be correct, but errors of exclusion through default, due to lack of evidence, would be far more extensive. Thus, insistence on direct hydrologic information is in effect a means of pretending that most real wetlands are not wetlands because we do not have hydrologic records to prove that they are. In fact, this whole approach could have been hatched by a criminal defense lawyer, to whom it would make perfect sense. In the case of criminal law, we insist on elaborate proofs of criminal guilt because we are unwilling to punish the innocent. For this reason, we either liberate or fail to prosecute swarms of criminals, but we accept error of exclusion as our price for sparing the innocent. The analogy with wetlands would only work under a national policy whereby we assume that all land is upland unless someone proves by use of water table data that it is not. Such a policy would leave many wetlands unprotected. Because regeneration of converted wetlands is unlikely, it makes most sense to minimize the conversion of wetlands through evidentiary errors, given that the national policy is *no net loss* of wetlands.

Another twist in the argument for independent evaluation of hydrology is that direct evaluation of hydrology, even when possible, is often less reliable than an indirect assessment of hydrology from soil and vegetation. For example, if we made a direct hydrologic assessment over a period of five years, which is quite a long time in the context of land development, we would still have only a sketchy idea about the recurrence interval for saturation. On the other hand, soils carry the record of hydrologic condition over centuries while vegetation gives us a good indication of hydrologic condition over the past 10 to 50 years (reflecting the growth of the largest plants). Thus in this sense, direct hydrologic information could be more confusing than helpful unless it is truly extraordinary in amount or quality.

A Necessary Evil

In some instances, the direct assessment of hydrology is unavoidable or helpful, even if it must be done rather poorly. These exceptions fall under three headings: (1) hydrologic change, (2) mixed evidence from other indicators, and (3) absence of other indicators.

Soil and vegetation may bear false witness to the presence of a wetland if the water source for the wetland has been removed or reduced. In such a case, the site is actually virgin upland, but the signs of its having been wetland still remain. Because humans have a great propensity to redirect water, the extinction of wetlands in this way is not at all uncommon. A similar change can even occur naturally, as it does when a river changes course or a beaver dam blocks a stream. Redirection of water flow may also create wetlands. Interestingly, those of us who are regulated moan pitifully about the federal rules that protect wetlands created accidentally by water diversion projects,[13] but there is less outcry about wetlands that become uplands when a natural water source is moved. If we take seriously the policy of *no net loss*, we should probably regard new wetlands incidentally created by water management as partial restitution for natural wetlands dried up by water management elsewhere.

If information on soil and vegetation fails to provide a clear diagnosis for a given site, hydrologic information may help resolve the ambiguity. This is a situation in which a year's data on water table depth could be used in conjunction with computer modeling based on weather records to indicate the status of a site. A more common path, and one that is likely to be acceptable unless it is applied by someone who is biased, is the use of professional judgment by a delineator who has many years of field experience in the region. Professional judgment in this field is anathema to some, but in fact is standard practice in virtually all branches of applied science.

Hydrologic assessment may be important or even essential where vegetation, soils, or both have been removed or recently disturbed. Agricultural lands often present this problem. For example, the entire landscape in large portions of the Mississippi floodplain consists of hydric soils because all of it was wetland during colonial times. Be-

---

13. Wetlands that are created by irrigation on farms typically are excluded from protection.

cause of the alteration of hydrology to accommodate farming and residences, however, one cannot read much into the presence of a hydric soil in these areas today. Where vegetation also has been removed for farming, there is no choice but to study hydrology and make a judgment on this basis. Fortunately for delineators, areas that were farmed prior to 1985 are not regulated as wetlands(unless abandoned agriculturally), even if they match the hydrologic profile of wetlands.

Direct assessment of hydrology will never be irrelevant to the identification of wetlands. It should be viewed, however, as a necessary evil to be accepted under special circumstances involving alteration, uncertain diagnosis, or removal of indirect indicators.

5

MOTHER EARTH

The virtue of soil is one of only a few axioms of American culture. Respect for soil, which can be instilled at an early age without rousing political or religious controversy, is the mark of agrarian origins. Nevertheless, or perhaps for this very reason, the average American knows more about asteroids than about soil. Even scientists, who are responsible for knowing pretty much everything, seem generally less aware of soil than of water, air, or even rock. Thus, this chapter requires a primer on the nature of soil.

The Fundamentals of Firmament

The USDA, which is the center of gravity for soil science in the United States, has defined soil as "earthy materials . . . on the earth's surface . . . capable of supporting plants out-of-doors."[1] To the thoughtful novice, this may seem a woefully unsatisfactory definition. In fact, one might feel downright indignant upon consulting a dictionary only to find that "earthy" means "having properties of soil." Thus, the USDA seems to be saying that soil is something that is like soil. This could be a simple case of cheating or perhaps a devilishly clever means of suppressing debate about the definition of soil. Fur-

---

1. Soil Survey Staff (1975). The USDA actually uses a number of definitions in its numerous publications on soils.

ther reflection, however, suggests that the reference to plants is more significant than it first appears to be. Plants do not grow on bare rock, shifting sand, or the bottoms of oceans, all of which are non-soil substrates. Thus, the USDA definition may say it all, but is masterfully understated in doing so.

Soil science textbooks are also a source of definitions, some of which are gratifyingly descriptive. Birkeland's (1984) textbook, which gives a geological perspective on soil, specifies that soil is a natural entity; is composed of mineral or organic constituents, or both; differs from the material from which it was originally derived; and reflects the operation of pedogenic processes. Pedogenic processes, in turn, include the addition of organic matter from plants and other substances from the atmosphere; removal of substances by water through erosion or leaching; and a variety of internal transformations and transfers involving organic matter and minerals. Birkeland also conveys the impression, as do other textbook authors, that the definition of soil is not a matter of high formality. In fact, most scientists have little interest in refining the definition of any general term such as soil (or wetland). Nuances of meaning become an issue only when such a term takes on legal and regulatory significance. For reasons explained below, the definition of soil has greater regulatory importance now than it did prior to the evolution of wetland regulation and may consequently require in the future more careful attention from soil scientists.

Soil scientists use hierarchical systems for classifying and grouping soils (Soil Survey Staff 1992). The presently dominant system, which is relatively new, divides soils into 11 orders. These orders are then divided into 47 suborders. Suborders are divided into great groups, and great groups are divided into subgroups. Subgroups are split into families, which are composed of series; a series may have a number of phases.

The name of a particular soil is made by concatenation of roots that correspond to various levels of classification. For this reason, the names of soils can be quite awe-inspiring (e.g., Plinthic Kandiaqualf) and, while logical, are scarcely an invitation for the uninitiated to learn more about soil science.

All 11 orders of soil contain at least some wetland soils. Among the 11 orders, however, only 1 order, the Histosols, is dominated by wetland soils. Histosols have high organic content and form under conditions of high moisture.

Soil scientists use, in addition to their classification, some general terms that describe the condition of soils. One such term is *aquic*, which refers to moisture regime. This term recently has been treated with great formality because it facilitates the description of wetland soils (hydric soils; Soil Survey Staff 1992). To the USDA Natural Resource Conservation Service (NRCS, formerly the SCS), an aquic moisture regime is one involving repeated saturation that is sufficiently prolonged to cause loss of free oxygen from the soil. Thus, a soil that is waterlogged for prolonged intervals without losing oxygen does not have an aquic moisture regime. The implications of this definition need further explanation, which will appear later in this chapter, but for present purposes it suffices to say that soil conditions are aquic only if the soil is wet more than the minimum amount of time required to cause the elimination of free oxygen.

Aquic moisture regimes cut across the soil classification system (i.e., any order of soils may be exposed to aquic conditions where drainage is poor). The aquic designation can be used for any soil except a Histosol. The reasoning, presumably, is that Histosols are known to be aquic, so we do not need to call them aquic.

Soils are described not only by moisture regime but also by temperature. Temperature has a bearing on accumulation of organic matter. If all other factors are equal, a warm soil typically will accumulate less organic matter than a cool soil (figure 5-1). This is explained by the tendency of organic matter to decompose more rapidly at higher temperatures (reference the temperature-metabolism rule of chapter 4). Thus, microbial decomposers are more likely to deplete the organic matter in a warm soil than in a cold one. The amount of organic matter in soil is relevant to the diagnosis of wetlands because organic matter supports the growth of microbes, which in turn cause depletion of oxygen from a soil that is saturated with water. Depletion of oxygen in soil is one of the most common conditions associated with wetlands.

Temperature, moisture, and organic matter of soils interact in ways that are understood in principle but difficult to predict specifically. For example, while a cool soil can accumulate organic matter more rapidly than a warm soil, the repeated saturation of a soil with water adds complications. Saturation leads to loss of oxygen; loss of oxygen reduces the rate of decomposition of organic matter. Thus, a soil that is frequently saturated with water may accumulate quite a bit of organic matter, even if it is warm (figure 5-1). Clearly, the diagnosis of wetlands through soils is not absolutely simple.

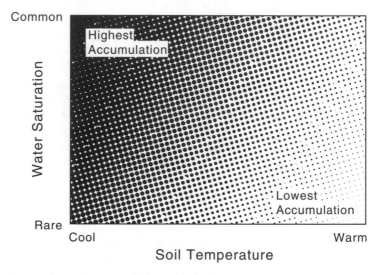

Figure 5-1. A diagrammatic view of the relationships among soil temperature, saturation of soil with water, and the rate of organic matter accumulation. Organic matter accumulates most rapidly in soils that are cool and saturated with water continuously or for extended intervals each year.

The process of soil development causes most soils to be layered. The layers, or horizons as they are called by soil scientists, are designated by letters (figure 5-2). The O horizon,[2] which consists of the superficial layer of litter derived from the plant canopy, typically is thin and has a consistency much like that of a compost heap. The A horizon corresponds more closely to what most people think of as soil. Unless removed by erosion, it varies in thickness from several inches to many feet but often falls in the range of a foot or two. The A horizon typically contains more organic matter than the layers below because it is closest to the source of organic matter (vegetation at the surface). Because the A horizon overlaps with the upper part of the root zone (ca. one foot), which is the critical zone for the recurrent, prolonged saturation that accompanies the development of wetlands, it is of particular interest to the identification of wetlands. The lower hori-

2. Some authorities refer to each major layer in the plural (e.g., the A horizons) on grounds that a major layer is made up of subsidiary layers, or horizons. For the uninitiated, it is simpler to use the term horizon as a collective noun to represent any one of the major layers bearing the letters O, A, B, or C.

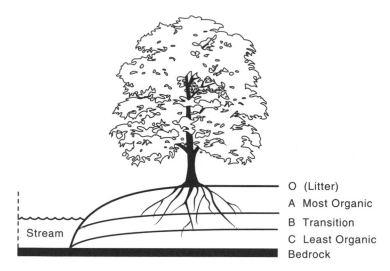

Figure 5-2.   Illustration of the layering of soil. The zone of greatest
relevance to the classification of wetlands is A, which supports extensive
root growth and holds most of the soil's inventory of organic matter.

zons, which are designated B (below A but strongly influenced by A)
and C (below B and less influenced by A), contain much less organic
material than the A horizon. Horizons can, of course, be further subdivided and classified.

Without having at least glanced at soil maps, one may think of different soil types as occupying huge blocks of land corresponding to
different climatic regimes of the United States. In truth, soils are about
as varied spatially as vegetation; in fact, the two often influence each
other. Thus, the landscape is a soil mosaic that often shows a great
deal of fine structure in the form of differing soil types interdigitated
in complex ways, even on a small farm (figure 5-3). The soil mosaic in
part reflects the influences of topography and drainage on soil formation, in part the irregularity in distribution of various kinds of soil par-
ent materials (rocks), and also the varied removal of surface layers
through erosion over long periods of time. The complexity of soil dis-
tributions is a source of complexity for wetland identification because
it is not possible to assume that the presence of a wetland soil at a
given spot indicates the presence of wetland soil throughout the surrounding area. Thus, a certain amount of mapping must accompany
any analysis of soils for the determination of wetland boundaries.

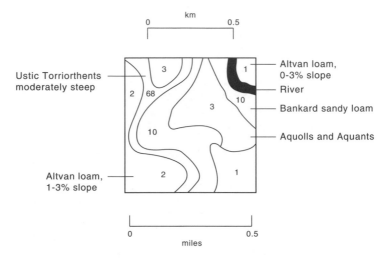

Figure 5-3. Soil map of a 160-acre block of land near the South Platte River in Colorado. Source: USDA Soil Conservation Service (1980).

Soils have been mapped in considerable detail throughout arable portions of the United States. The mapping process, which is the responsibility of the USDA's NRCS, has been underway for decades (Simonson 1997). The maps were intended for agricultural planning but can be quite helpful in showing the location of wetland soils. Even so, changes in drainage, unavoidable inaccuracies and simplifications in mapping, and lack of specific interest in wetland soils prior to about 1975, dictate that general soil maps (soil surveys) are only suggestive and not definitive for mapping wetlands.

Redox and Related Arcana

One of the few unsullied dividends of the nuclear age is that most of us know a little something about atoms. All except the chemically impaired will remember, as a result of having been taught three times (grade school, high school, and college), that atoms are composed of a nucleus surrounded by electrons, and that they sometimes share their electrons with, or lose their electrons entirely to, other atoms. This fleck of knowledge is sufficient to support a rudimentary understanding of redox.

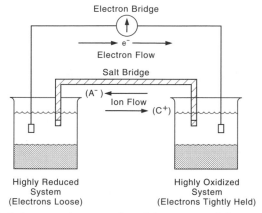

Figure 5-4. Laboratory demonstration of the passage of electrons from a more reduced system to a less reduced system.

Redox is the economy of electrons. When an atom takes up an electron, it is said to be reduced. When it gives up an electron, it is said to be oxidized. Thus, reactions involving the exchange of electrons are commonly called reduction–oxidation reactions, or redox reactions. Redox reactions occur in the atmosphere, soil, water, and in combinations of these, and thus in wetlands.

In water solutions or wet sediments, electrons are attached to an atom or atom cluster; they are not completely free. The electrons in such a system vary, however, in the firmness with which they are bound. We may think of a highly reduced system as being rich in loose electrons, that is, some of the electrons could be passed easily to other atoms or atom clusters that have positive charges if such electron acceptors were made available. In contrast, even the loosest electrons of a highly oxidized system are difficult to detach.

As shown in figure 5-4, a highly reduced system will pass electrons to a highly oxidized system if the two are connected by a pair of bridges. The first bridge, which is for electrons, consists of a wire that will accommodate the movement of electrons. The second bridge is filled with a watery gel containing a sufficient amount of salt to make electrical contact between the two systems. When the two bridges are in place, electrons will flow spontaneously from the reduced system to the oxidized system. At the same time, following nature's tendency to keep the number of positive and negative charges equal, ions migrate over the salt bridge.

The electron exchange contrived between beakers under laboratory conditions can be converted to a measurement system that is expressed by a scheme very similar to the one that is used for temperature. A redox measurement, which is called the *redox potential*, is a value on a scale consisting of a reference point with a value of zero and positive and negative values extending to either side of the reference point.

For redox, the reference is the balance point for the hydrogen electrode: a system has a redox potential of zero if it will neither donate electrons to nor accept electrons from a hydrogen electrode. Given that a discussion of hydrogen electrodes can be tiresome at best, it suffices to say that a hydrogen electrode consists of a mixture of water, oxygen, and hydroxide ions that can either receive or donate electrons. The direction and force with which electrons flow to or from such an electrode (or any other kind of electrode that can be calibrated against it) provides a means of measuring redox potential. The measurement is expressed in terms of millivolts; the volt, of course, is the standard measure of tendency for electrons to flow.

Highly oxidized locations in the environment, such as an aerated soil, have redox potentials of +500 millivolts or higher (figure 5-5). The most highly reduced locations in the environment, such as wet organic muck, have redox potentials of -500 millivolts or lower. Natural environments encompass the entire range between these extremes. In fact, a given ecosystem may show very different redox potentials at different locations (e.g., the surface water of a lake vs. its sediments).

The nagging question at this point is why redox potentials should vary from one place to another. The wide divergence in redox potentials across and within ecosystems is caused mainly by two factors: (1) the metabolism of plants, animals, and microbes and (2) the movement of oxygen gas (free oxygen, or O_2).

Oxygen holds electrons tightly. The addition of free oxygen to a system tends to raise the redox potential of the system because oxygen is reluctant to give up electrons. Removal of free oxygen lowers the redox potential because few other substances hold electrons so tightly. Organisms harness this principle in their internal controls over reduction and oxidation: they dump oxygen when conducting reduction and take up oxygen when conducting oxidation. Biological uptake of oxy-

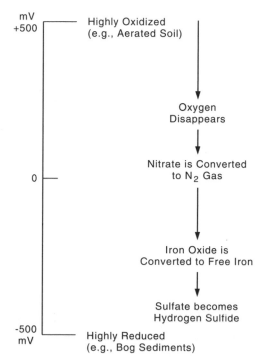

Figure 5-5. Changes that occur as the redox potential moves from high to low. For more detail, see Reddy and D'Angelo (1994).

gen can be sufficient in some cases to deplete an environment of oxygen. A location that lacks free oxygen is called anoxic; only anaerobic processes can occur there.[3]

The two most important metabolic processes influencing redox potential are photosynthesis and respiration. Photosynthesis involves the uptake of electrons from the environment as necessary for the reduction of carbon, which leads to the synthesis of organic matter. Thus a photosynthetic organism carries out an internal reducing process (photosynthesis) during which it dumps oxygen, which oxidizes the external environment. Respiration is the reverse: the organism imports oxy-

3. There are some variations of opinion on appropriate use of the terms anoxic and anaerobic. For present purposes, soil or water lacking free oxygen will be called anoxic. Processes that occur in an anoxic environment will be called anaerobic (cf. Brady and Weil 1996).

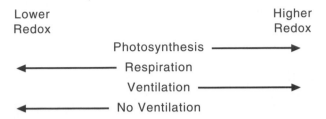

Figure 5-6. A view of factors that promote increase or decrease in redox potential.

gen, which allows it to carry out internal oxidation to produce energy. The external environment, from which oxygen is removed, is correspondingly reduced (figure 5-6). Where photosynthesis predominates over respiration, the environment will move toward a high redox potential and have a high amount of free oxygen. Where respiration predominates over photosynthesis, the environment will move toward low redox potential and free oxygen will eventually disappear.

Movement of free oxygen from the atmosphere may influence redox potential. If mixing with or diffusing into a location that is moving toward low redox potential, free oxygen will raise the redox potential (figure 5-6). Thus, free oxygen from the atmosphere can offset or reverse the tendency of respiration to lower the redox potential of soil or water.

Given the influences of photosynthesis, respiration, and movement of free oxygen, one can predict the general pattern of redox potentials in the environment. Where photosynthesis is occurring, the redox potential is typically high. The upper zones of lakes and oceans, where there is sufficient light to sustain photosynthesis, are examples. Where there is no light for photosynthesis, respiration will predominate and redox potentials will tend to be low, unless the continuous penetration of free oxygen offsets the effect of respiration on the redox potential. For example, a sediment surface in the depths of a lake often has a low redox potential whenever the mixing of the lake does not extend to the bottom. Photosynthesis is not possible there due to lack of light, and there is no mixing to bring fresh oxygen from above. As respiration uses up the oxygen, the redox potential declines. When the lake mixes, however, the redox potential rises as free oxygen again reaches the sediment surface.

The Ventilation of Soils

Soils contain not only mineral material in the form of clay, silt, sand, and coarse bits of weathered rock, but also organic matter that originates mostly from plants. Much of this organic matter has been extensively altered by biological action. The organic matter of soil serves as food for microbes and other organisms that eat the organic matter, the microbes, or both. All of these organisms respire, as do the roots of plants. Thus, soil often supports much respiration.

Photosynthesis cannot occur within soil because light does not penetrate soil more than a few millimeters. Therefore, respiration is the consistently dominant metabolic process in soils. The basic facts about redox thus lead us to expect that soil will tend to develop a low redox potential because of the predominance of respiration, unless this tendency is offset by continuous penetration of free oxygen.

Soils show varying degrees of ventilation from above. Because all soils consist of particles, they all have considerable space between particles (Brady and Weil 1996). As long as this space is not completely filled with water, a certain amount of gas exchange will occur between the soil and the air above it. In other words, soils are ventilated unless they are waterlogged. The degree of ventilation, of course, varies greatly with the size and number of pores in a soil.

Soils that are open to ventilation typically receive enough oxygen to offset the demands of respiration, although the amount of free oxygen may be less than in the air above. Thus when soils are not saturated with water, they typically have moderate to high redox potentials. Saturation of soil with water, however, shuts off ventilation by filling the pore spaces so that free exchange of gas between the soil pores and the overlying air is blocked.[4]

When soils first become saturated with water, a certain amount of oxygen remains available because the water itself contains a few parts per million of free oxygen in dissolved form. Thus for a time, the blockage of ventilation may have little effect on the redox potential of the soil or on metabolic processes of soil organisms. The dissolved oxygen is not replaced, however, because water has blocked exposure of the soil to air. Respiration eventually removes all free oxygen from

---

4. In theory, oxygen gas could penetrate even a waterlogged soil by dissolving in the water of the uppermost pores and then diffusing downward. As a practical matter, this mode of gas penetration is so slow that it need not be considered.

the soil. As the soil becomes anoxic, it undergoes drastic biotic and chemical reorganization, much like a bankrupt company as it runs out of cash.

Most kinds of organisms require oxygen for respiration. When oxygen disappears, they move, die, or enter a dormant state in which they do not respire much. Invertebrates such as insect larvae and earthworms living in soil would respond in these ways as oxygen disappears from the soil. Microbes, however, are a different story.[5]

Some kinds of microbes, especially bacteria, can respire in the absence of oxygen (Schlegel 1993). Respiration requires the transfer of electrons onto an electron acceptor. While most kinds of organisms must use oxygen as the electron acceptor, microbes often can use other kinds of atoms as electron acceptors. For example, nitrogen in soil often takes the form of nitrate (NO_3^-). Nitrogen in this form can accept five electrons per atom, and in the process becomes free nitrogen gas (N_2). This process (denitrification) can be facilitated by a wide variety of bacteria, any of which can sustain their respiration in this way. Iron, manganese, sulfur, and carbon also can take up electrons through the metabolic manipulations of bacteria. Thus, bacterial respiration continues long after oxygen has disappeared (figure 5-5). As respiration continues, the redox potential of the soil declines progressively.

Lifestyles in an Anoxic Soil

Anoxic soils exclude organisms that must use free oxygen from soil pore spaces. Woody plants, herbs, and grasses are an interesting case in this regard because their roots, which may comprise as much as half of their total mass, are exposed to the anoxic environment beneath the soil while their stems and leaves are open to the atmosphere. Except for the discomfort of being buried waist deep, a human would have no difficulty in this situation because the oxygen essential to support respiration of the entire organism can be taken in at the top and pumped throughout the body by an efficient circulatory system. Plants also have circulatory systems, but they are less efficient at moving things from top to bottom than from bottom to top. For most plants, the oxy-

5. A few kinds of organisms other than microbes can function in the absence of oxygen (Fenchel and Finlay 1995).

gen in a leaf cannot be conveyed efficiently to the tip of a root. Thus, most plants die when their roots are exposed to anaerobic soil for a prolonged interval. This explains why we frequently see bald spots on wet areas in a large cornfield, and why farmers often want to drain wetlands.

While many plants are very sensitive to loss of oxygen from soil, some can withstand prolonged absence of oxygen in the root zone. Plants that survive well or even thrive in anoxic soils are called hydrophytes. They possess various adaptations that compensate for lack of oxygen in the soil. These will be revealed more fully in the next chapter, but one example is the extension of roots above the soil surface in the form of a pneumatophore, as seen in mangrove swamps. A less visible but even more important example is pressurized ventilation, which is a way of moving gas down to the roots without a pump.[6]

Challenges to plants in saturated soils do not always end with depletion of oxygen. Because the anaerobic lifestyle of microbes in saturated soils involves the transfer of electrons to a variety of electron acceptors, continued decline in the redox potential after the depletion of oxygen results in a cascade of chemical changes. While the details are sufficiently complex to provide employment for a horde of soil scientists and plant physiologists, the main biotic effects of chemical changes in anaerobic soils can be understood in terms of two simple principles: (1) the decline of redox potential in soils is accompanied by conversion of oxidized chemical forms to reduced chemical forms (figure 5-5) and (2) reduced chemical forms are often toxic to organisms. For example, a highly reducing environment will produce sulfide, which is a reduced form of sulfur that otherwise would be present as sulfate. Sulfate is harmless to organisms, but sulfide is highly toxic. Thus, the accumulation of sulfide and other reduced chemical substances increasingly toxifies the soil as the redox potential declines.

The toxic substances accumulating under increasingly low redox potentials create a progressively hostile environment for most organisms. Plants that might tolerate mere loss of oxygen in the root zone may be eliminated by toxicity. Only plant species specifically adapted to these chemical stresses in the soil survive the chemical conditions that accompany the lowest redox potentials.

6. Pressurized ventilation comprises a cluster of related systems that harness differences in temperature or humidity to achieve a one-way flow of gas (leaf to root) or a circular flow of gas (leaf to root and root to leaf). See *Aquatic Botany,* Vol. 54, 1996.

Reading the Redox

The redox potential of a soil can be measured with a probe. Such measurements are not especially useful, however, unless they are made frequently over an extended period of time. Because redox changes from week to week and season to season, and even varies between years, a single redox reading for characterizing the redox condition of a soil is about as useful as a single air temperature reading for characterizing the climate of a state. Furthermore, neither the United States nor any other country collects redox data from soils on a routine basis except in a few special study sites. Thus, information about the redox status of soils must come from some source other than measurement of redox. Fortunately, soils that repeatedly reach low redox potentials often show clear physical indicators of reduction. These indicators are called *redoximorphic features*.[7]

Most soils that develop low redox potentials do so only intermittently. For example, the soil of a prairie pothole or riparian swamp may be fully saturated with water for weeks or months when water is most abundant. During most of this interval of saturation, the soil will be anoxic. With the onset of dry weather, however, the pore space may again be open to ventilation, which returns the soil to an oxidized state. Thus, any signs of anoxia in the soil must be persistent through periods of soil oxidation in order to be useful in the analysis of soil. Redoximorphic features are slow to form but once formed are not eliminated by intermittent periods of oxidation. Thus, they are ideal indicators of the recurrence of reduced conditions because they will be visible even if the soil is not anoxic at the time it is examined.

One of the most valuable redoximorphic features is mottling. Mottling is produced by repeated reduction and oxidation of iron in the soil. Under oxidized conditions, iron and manganese are highly insoluble. They are distributed through the soil in combined form as oxides or hydroxides. When the redox potential of a soil declines to about 0 millivolts (figure 5-5), conversion of oxidized iron (Fe^{+3}) to its re-

---

7. The four types of redoximorphic features recognized by the NRCS are as follows: (1) redox concentrations (zones of accumulation for iron and manganese oxides, marked by irregularities in soil color); (2) redox depletions (zones from which iron and manganese oxides have been lost, as marked by color variation or other indicators); (3) reduced matrix (darkness of soil color undergoing a change to greater brightness after exposure to air); and (4) presence of free iron (as indicated by a chemical test). These are given in the SCS keys to soil taxonomy (Simonson 1997).

duced form (Fe^{+2}) begins under the influence of microbes.[8] Reduced iron, unlike oxidized iron, is highly soluble. Thus, the iron that was previously bound to particles in the soil in solid form is liberated to the water that fills the pores of the soil. The distribution of reduced iron, however, is not necessarily uniform; it may accumulate in small patches of the soil where the redox potential is lowest or where textural irregularities in the soil affect its liberation or movement. When a reduced soil containing free iron begins to dry, the redox potential rises as oxygen enters and iron is reoxidized, again with the participation of microbes. Because the dissolved iron is distributed irregularly, the precipitation process is also irregular.

Oxidized iron has color. The lesson is clear in any washbasin or around any pump served by water containing some reduced iron: the reduced iron in the water oxidizes as it sits on an open-air surface and makes an orange stain. Similarly, the precipitation of iron in soil imparts color to the soil. The color may or may not be strictly orange because it may mix with other colors in the soil. Uneven precipitation of iron will, however, cause an uneven distribution of color, or mottling, in the soil. This explains the significance of mottling as a redoximorphic feature of soils.

Another characteristic of reduced soils is dark color. Darkness in a hydric soil is accounted for by organic matter and by reduced substances, which as a rule are darker than oxidized substances. The presence of darkening agents (organic matter and reduced substances) mutes the background color of the soil, which is determined by its mineral composition. For any given color, the degree of muting is referred to as the chroma. Soils in which the background color is sharply developed are said to have high chroma; those having colors that are muted by the presence of darkening agents are said to have low chroma. Thus, the soils of wetlands have low chroma.

The properly equipped soil scientist carries a little booklet called a Munsell color chart. The booklet, which allows visual determination of chroma (figure 5-7), provides a diagnosis of the redox history of the soil. Even though this may seem a bit like tarot, it makes perfect sense in view of the association between reducing conditions and the appearance of soil. The user of the chart first finds the page containing the appropriate hue. This page will show varying degrees of darkening

8. Mottling may involve not only iron but also manganese, which is reduced from Mn^{+4} to Mn^{+2} as redox potential declines.

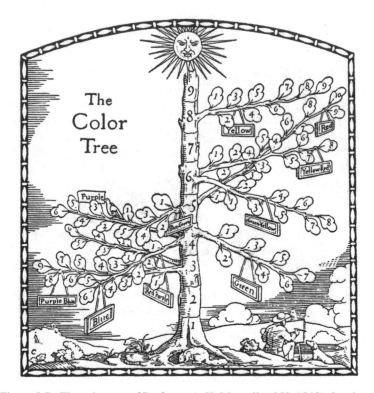

Figure 5-7. The color tree of Professor A. H. Munsell (1858–1918) showing the principles upon which the Munsell color chart is based. The hues (e.g., red and yellow) are arranged radially around the trunk. Value, which is the degree of brightness on a scale from white to black, is indicated by the position along the vertical extent of the trunk. Chroma, which is the strength of the color, varies from highest (purest) at the horizontal points most distant from the trunk, to lowest (darkest or grayest) at points closest to the trunk (Birren 1969).

(graying) for the hue (e.g., maroon is a red of low chroma and vermillion is a red of high chroma). Strong darkening, as shown by a match of the booklet colors with the soil in the field, indicate low chroma, and thus, a history of low redox potential.[9]

9. The visual diagnosis of chroma is not infallible. Some intrinsically dark minerals found in soil can be mistaken for the darkening associated with low redox conditions.

The Vexing Matter of Duration

Anoxia in saturated soils produces many of the distinctive features that we associate with wetlands. Thus, we must ask how long it takes for oxygen to disappear from a waterlogged soil. This question highlights the unfair advantage of physicists over environmental scientists. If asked how rapidly a photon travels through space, a physicist can respond—smugly—that photons move at 3 x 10^8 m/s. If asked how long it takes for a saturated soil to become anoxic, a wetland scientist must respond—evasively—that it all depends.

The elapsed time between initial water saturation and first occurrence of anoxia in a saturated soil, while variable, is not entirely unpredictable. Las Vegas could make odds on it, for example, if the concerned parties were more inclined to bet than to litigate. A bookmaker would first want to know the rate of respiration, which is controlled mainly by the soil temperature and amount of organic matter. Basic principles provide some help here. The rule of doubling (figure 4-2) allows one to approximate the difference in oxygen depletion rate when a soil is 5°C as compared to the depletion rate in the same soil at 15°C. Comparison among soils is less predictable, however, because of the influence of organic matter. We can measure the amount of organic matter in a soil but not all organic matter is equal. A thick New York strip and a five-inch piece of 2 X 4, for example, have about the same amount of organic matter but scarcely offer the same nutritional prospects for a human being. Microbes, although more flexible than humans in this respect, have their limits and cannot make equal use of all organic matter. Thus, both the amount of organic matter and its nutritional value to microbes influence the rate at which oxygen disappears from the soil.

Temperature and the amount and type of organic matter are the most important influences on rate of disappearance of oxygen from soil, but other factors may introduce variation as well. For example, soils that appear to be waterlogged may contain some air spaces. The volume of these air spaces will affect the amount of time that oxygen is available in the soil. In other cases, water in the soil may move and the incoming water may bring oxygen with it. Thus, the problem of oxygen depletion can be described as largely a matter of temperature and organic matter, but with complications sometimes added by other factors.

The issue of duration (i.e., the amount of time that a soil must remain saturated in order to become anoxic) has proven problematic to those who need to link hydrology with soil typology. The amount of field evidence on this subject is scandalously small and provides only a weak foothold for generalization. When soils are cold, they may hold oxygen for weeks or months when fully saturated. When they are warm, they may lose their oxygen in only a few days.

A Soil Is Born

In its early search for practical ways of identifying and mapping wetlands, the Army Corps identified soil as one of the three major sources of information (the other two are water and vegetation) that could produce a reliable diagnosis of wetland. Soil scientists were, of course, well aware of the role of recurrent, prolonged saturation (aquic conditions) in producing low chroma and other redoximorphic features. Because such soils could be found in all soil orders, however, there was no easy correspondence between the soil classification system and the soils of wetlands.

The federal agencies grasped the nettle of wetland soil typology in 1977 through the establishment of a list of wetland soils.[10] This effort gave rise to the NRCS National Technical Committee on Hydric Soils (NTCHS) in 1981. At first, the NTCHS included mostly government soil scientists from the eastern and midwestern United States, but it subsequently became more diversified in its institutional and geographic affiliations. The NTCHS, which was formalized in 1985, defined and identified a new group of soils, which it called *hydric soils*, that develop under conditions of repeated and prolonged saturation. Hydric soil was intended as a new kind of classification that would cut across the taxonomic categories of soil for the practical purpose of facilitating the identification of wetlands.

One of the first jobs of the NTCHS was to define hydric soil. Although the NTCHS definition has undergone two transformations, all three forms of the definition have had two key features: (1) saturation of the soil with water at or near the surface for an extended interval and (2) development of low redox potential in the upper part of the

---

10. This account of the formalization of hydric soils is taken mainly from the National Research Council wetlands report and from personal communication with Dr. C. Johnston of the University of Minnesota, Duluth.

soil. The two components of the definition may seem redundant, given that low redox potential is unlikely in the absence of water saturation. Low redox potential is the real essence of the definition because a saturated soil cannot be hydric if it has a high redox potential, regardless of the duration of saturation. Saturation, however, is the only link between the definition and standard methods of soil classification: soil classification takes water saturation into account, but not redox potential.[11]

The second job of the NTCHS was to apply its definition of hydric soil to the 18,000 or so soils that are captured in the NRCS database at Iowa State University in Ames, Iowa. Had the data on each of these soils included redox potential, the application of the hydric soil definition to the list of soils, and thence to soil maps, would have been mostly a matter for clerks and computer programmers. Unfortunately, there is no information on redox or oxygen for most soils. Making the best of a difficult situation, however, NTCHS patched together a list of rules, which it called criteria, to be applied to the database. The criteria are of three types: (1) taxonomic, (2) hydrologic, and (3) hybrid. The taxonomic approach was used for Histosols, because Histosols form only under extremely moist conditions that are consistent with the definition of hydric soil. Thus, a soil that meets the taxonomic qualifications of a Histosol is automatically declared a hydric soil.[12] Beyond this taxonomic approach, the NTCHS allows any soil that is routinely ponded or flooded for considerable intervals during the growing season to be designated hydric on this basis alone. NTCHS also hybridized hydrologic and taxonomic approaches by designating as hydric certain suborders, subgroups, and great groups[13] showing water tables close to the surface (normally within 0.5 feet) for two or more weeks during the growing season, or by other similar means too specific to mention here.

The application of NTCHS criteria to the master list of soils allowed all soils in the classification system to be categorized either as

11. This is not an oversight, but rather a reflection of technical problems: reliable measurement of redox potential is difficult.

12. One small group of the Histosols (Folists) is systematically excluded from the hydric soil category because it does not form under wetland conditions. Changes in drainage are also taken into account (i.e., a Histosol that has been drained by natural or artificial means would not be called a hydric soil).

13. These are taxonomic units in the soil classification system (see the first part of this chapter).

hydric or non-hydric. With wise regard for potential controversy, the NTCHS left open the possibility for reclassification of any particular soil on the basis of petition involving evidence not available through the NRCS database. Predictably, reclassifications have tended to remove soils from the hydric category rather than the reverse. Perhaps the most prominent example is a reclassification removing 13% of the state of Florida from hydric soil classification (Hurt and Puckett 1992). Because the criteria for identification of hydric soil are inferential, their modification for the reclassification of individual soils presents at least the possibility of circular logic, even with the best of intentions.

The selection of criteria for hydric soils has served the urgent need of the NRCS to map wetlands by some reasonably easy method that can be applied over large areas. Ironically, the most important criterion for mapping wetland soils is purely hydrologic: the NRCS has used the two-week flooding or ponding features of the hydric soils criteria to map wetlands on agricultural lands (chapter 4).

Hydric Soil as Dogma

Hydric soil, although only recently defined, has with amazing speed assumed a respected niche in the agricultural and soil science communities under the guiding hand of NRCS and other federal agencies charged with the identification of wetlands. In fact, there are no practical alternatives on many agricultural lands. Rigorous consideration of hydrology, which would require data on subsurface saturation, is impractical. Plant community analysis, which is the most common means by which wetlands are identified on nonagricultural lands, is often impossible because agricultural lands can be without vegetation or show extreme disturbance of natural vegetation. Thus the agricultural community, under the leadership of NRCS, increasingly has identified wetlands with hydric soils.

The commitment to hydric soils may have gone too far. For example, the 1985 Food Security Act, which offers one of the two existing federal legislative definitions of wetlands, defines wetlands around hydric soils. Ecologists, who were agreeably nodding their approval at the mobilization of soil science around the problem of identifying wetland soils, took alarm when it became clear that hydric soil had begun to assume the status of universal currency for wetland identification on agricultural lands.

There are several fallacies in the assumption that wetlands must have hydric soils. First, wetlands can be defined quite well without any reference to hydric soil. Second, the concept of hydric soil, although admirably well developed by the NTCHS, is still inherently fuzzy and subject to considerable interpretation. Third, the mapping of hydric soil in practice is often done by the evaluation of surface hydrology rather than soil itself.

The National Research Council's reference definition of wetlands (chapter 2) mentions hydric soil is as a common indicator of wetland, but not as a requirement. Soils that form in wetlands retain some hydric features after they have been drained by natural or anthropogenic means. This is a troublesome aspect of hydric soil but is not especially controversial because drained soils can be treated as a special case. The controversy centers around the reverse case, i.e., the existence of wetlands where hydric soils are absent. Most agriculturally oriented authorities do not acknowledge this possibility, whereas most ecologically oriented authorities do.

There are at least three conditions under which wetlands can form where hydric soils are absent: (1) prolonged saturation of a substrate not defined as soil; (2) prolonged saturation of soil in the presence of oxygen; and (3) prolonged saturation leading to anoxia but not to chemical reduction of iron.

Geologists are as a group very liberal in their definition of soil: to most geologists, any surface that can be scuffed with the toe of a boot is soil. Soil scientists have been more fastidious—probably because they are charged with naming and mapping soils, which geologists are not. Soil scientists may, for example, exclude loose sand and substrates that are perennially inundated. In fact, the idea of agricultural utility has influenced the definition and classification of soil in the United States. This introduces a concern for proper recognition of wetlands through the identification of soil types: wetlands need not have actual or potential agricultural utility. Thus, the definition of wetland cannot be made subservient to the definition of soil as long as the definition of soil reflects the concept of agricultural utility.

There is no need for the definition of wetland to be restricted by the definition of soil. For example, a soil scientist may not consider a loose alluvial substrate in a floodplain to be a soil, particularly if such substrate had been disturbed by flood so recently that no mature plant community had yet become established on it. If this substrate were recurrently saturated at or near the surface for long intervals, however, it would likely support a wetland. At some point, probably coincident with the maturation of the wetland plant community, the soil scientist might be ready to recognize the substrate as a soil. The existence of the wetland, however, would not be conditional on the classification of the substrate as a soil. Thus, wetlands could be found where a soil scientist sees no soil at all.

Another possibility is the presence of a wetland where the substrate is saturated with water for extended intervals but never loses its oxygen. Such a situation could occur, for example, where oxygenated water is moving through a sandy substrate to such an extent that the consumption of oxygen is balanced by continual resupply of oxygenated water. Another possibility would be the absence of sufficient organic matter in a soil to allow complete depletion of oxygen. Yet a third possibility is that the substrate remains sufficiently cold during the growing season (e.g., at high latitudes or high altitudes) to prevent respiration from using all the oxygen in the substrate.

Some authorities on soil and hydrology view the foregoing arguments as special pleading by ecologists who are interested in any sort of revisionist thinking that makes wetland boundaries cover more area. Indeed, these are just the sorts of arguments that would appeal to someone who wants to counter the argument of the agriculturalist that a wetland must have a hydric soil. In reality, however, the possibility of wetlands in oxygenated substrates derives from the definition of wetlands and not from environmental sophistry.

Wetlands must have ecologically distinctive properties that are traceable to recurrent and prolonged saturation. The question for oxygenated substrates is whether the presence of water very near the surface or at the surface for prolonged intervals can induce significant ecological responses if the water in the soil retains at least some dissolved oxygen. This question has been surprisingly little studied. It may not be a common occurrence, or possibly individuals on both sides of the underlying argument feel that they know the truth of the matter without even studying it.

The prolonged presence of water can induce biological responses that are not necessarily dependent on loss of oxygen. For example, trees that grow on perennially wet soils often develop buttresses because an unbuttressed trunk may be insufficiently anchored to hold the tree upright during prolonged intervals of soil saturation; this has no connection to oxygen in the soil. We can also easily see that many kinds of organisms ranging from mammals to invertebrates cannot complete their life cycles in a soil that is inundated or saturated, whether or not it has oxygen. Thus, there are good reasons to expect that a substrate subject to prolonged inundation or saturation at or near the surface would have distinctive ecological characteristics, even if it retained its oxygen.

Hydric soils are identified through low redox potential in the upper zone. Low redox potential in the upper zone in turn is deduced from redoximorphic features. These features, however, are most evident only where the redox potential crosses the iron threshold at about 0 millivolts. Oxygen may be absent at a redox potential of +300 millivolts. Thus, there is a gap between 0 millivolts and +300 millivolts within which oxygen will be absent but iron will not be liberated in reduced form.[14] Redoximorphic features will not form or will be weak under these conditions. Even so, the absence of oxygen may be a stress to many species of plants and soil organisms that require some oxygen in the soil. Thus, wetlands probably exist where oxygen is depleted from the soil but the redox potential is above the threshold for formation of clear redoximorphic features.

Some fierce arguments may develop between the ecologist or plant scientist who maintains that a wetland may develop where there is no hydric soil and the agricultural scientist or soil scientist who maintains that hydric soil is an essential characteristic of a wetland. A neutral observer might wonder if the argument has any great practical significance. Most wetlands have soils that easily meet the NTCHS definition of hydric soil. The residual wetlands probably account for only a small area and are relatively few over the United States as a whole; however, the argument may be important regionally. Biotic communities that have developed in response to prolonged saturation without anoxia or with anoxia but without a redox potential below 0 millivolts are likely to coincide with specific sets of geomorphic and climatic conditions. As mentioned above, these may include topography leading to transfer of oxygenated water underground, high substrate porosity, low organic matter content in the substrate, or low temperature during the growing season. These conditions will not occur at random

---

14. Manganese is liberated before iron but traces of its liberation and re-precipitation are not as easily seen as those of iron.

but rather in geographic clusters. Thus, the exclusion of wetlands showing these conditions, while not necessarily significant as a percentage of the national wetland inventory, could be quite significant regionally.

Even more important is the maintenance of objectivity and consistency in the identification and mapping of wetlands. Not all ecosystems that show distinctive characteristics reflective of prolonged and recurrent inundation have hydric soils. Furthermore, the concept of hydric soil is still evolving and is presently defined in the practical sense by a patchwork of criteria that are only partly satisfactory. Finally, there is danger that segments of the agricultural community may view negotiation over the classification of hydric soils as a means of summarily excluding some wetlands from the protection that was intended for them by law. The NRCS is still feeling its way toward holding a firm line on a technical matter (definition of hydric soil) against the headlong force of economic interests in drainage of wetlands.

Soil in Perspective

Soil is immensely useful for the mapping and identification of wetlands because it holds a cumulative record of soil saturation or inundation over a long period of time wherever saturation has been accompanied by low redox potential. Soil can give false positive evidence of wetland, however, where conditions necessary for inundation or saturation of soil have changed. In addition, wetlands may be present where the soil carries no record of low redox potential.

NRCS has done wetland science and wetland regulation a service in sponsoring efforts to define and list hydric soils in the United States. The emphasis on hydric soils can easily run to excess, however, if hydric soils are taken as a definitional requirement for wetlands. A wetland is an ecosystem that develops in response to the perennially prolonged presence of water. A hydric soil is a common, but not universal, symptom of the prolonged and recurrent presence of water.

6

THE CAST OF CHARACTERS

Through the inexorable workings of natural selection, wetlands have come to support a group of species that are especially well adapted to the physical and chemical peculiarities of saturated substrates and shallow water. These species give wetlands their distinctive biological signature and sustain the biotic functions of wetlands (Tiner 1998).

The species that occupy wetlands show varying degrees of specialization. Some are so highly specialized that they can live or reproduce only in wetlands; these are called *obligate* wetland species. A second group contains organisms that are adapted for life in wetlands but are not restricted to wetlands; these are called *facultative* wetland species. Facultative wetland species show every conceivable shade of association with wetlands: they range from almost obligate to barely facultative.

The distribution of obligate wetland species coincides closely with the distribution of wetlands. Thus, one could be tempted to rely heavily on obligate wetland species to find and map wetlands. Inference from obligate wetland species, however, must be tempered with caution. First, the absence of obligate wetland species does not necessarily mean absence of wetland, given that a wetland can be mostly or even entirely occupied by facultative wetland species. In addition, the degree to which a given species is a wetland obligate may not be known with absolute certainty. Where a genetic variant or an unusual set of physical conditions prevails, a species that seemed to be obligate in other situations may prove to be merely facultative and thus not diagnostic proof of the presence of wetland.

The perils of absolute reliance on obligate wetland species have turned the attention of wetland mappers to the analysis of entire communities. Although community analysis can be done in a number of ways, the central idea is to score a community according to the proportionate representation of species that show a known facultative or obligate affinity with wetland conditions.

The analysis of communities for the purpose of mapping and identifying wetlands has been focused almost entirely on vascular plants (grasses, forbs, shrubs, and trees). Two good reasons for this are the relative ease with which plant communities can be analyzed and the immobility of plants. Plants can be observed and tallied by transect sampling methods. While the identification of plant species requires much more than a warm feeling about nature, it can be readily achieved through training and experience. The immobility of plants ensures that their presence signifies the suitability of conditions for their growth at the spot where they are found. Thus, while vascular plants are not the only possible choice for community analysis, they are certainly the most practical and most informative in most cases.

The Spice of Life

Nature promotes diversity spontaneously. A cornfield, which if well maintained is an extreme example of low diversity,[1] can be sustained only by a combination of ingenuity, effort, and regular infusions of energy. Only a few natural communities lack diversity, and these occupy the most extreme environments on the planet.[2]

The most obvious way to measure the diversity of a community would be to count the number of species in it. This seemingly simple task presents an amazing number of problems. Most of the individuals in a community belong to a few common species, whereas most of the species in the community are represented by small numbers of individuals (i.e., they are rare). Thus, increasingly rigorous examination of

1. A community that lacks species diversity entirely would consist of a single species.
2. For example, Iltis (1975) found that hypersaline lakes of North Africa generally contained only a single species of algae, whereas the algal communities in shallow lakes of low to moderate salinity would generally be home to 100 or more species.

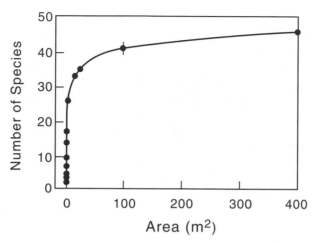

Figure 6-1. Species–area curve for plants in British bog (Hopkins 1957).

a community merely turns up progressively rarer species. Unless the community is very small and each individual in it can be examined, the enumeration of increasingly rare species becomes a job of finding needles in larger and larger haystacks. Therefore, a complete enumeration of the species in communities is not a feasible method for establishing community diversity.

The relative abundances of species in a given community are easily summarized by a species–area curve (figure 6-1). Such a curve can be constructed with information from repeated sampling. For example, a plant ecologist might use a standard sampling area of several square meters. A location for sampling would be determined at random and the species within it would be listed. This single sample would be likely to contain several of the most common species and might also contain one or more less common species. A second sample would then be located and analyzed in the same way. The second sample would probably contain many, but not all, of the same species as the first sample, and perhaps one or two species that were not in the first sample. A large set of such samples could support construction of a graph showing the growth in cumulative number of species against cumulative area sampled. This is the species–area curve (Rosenzweig 1995).

Species–area curves increase steeply at first and then begin to flatten as sampling uncovers rarer and rarer species. The curves for different kinds of communities may rise at different rates and flatten at different points. Communities having curves that rise slowly to great heights require the greatest effort to characterize.

Another way to capture some of the differences in species diversity among communities is by reference to the number of species per thousand randomly sampled individuals. For example, a boreal forest would show approximately 5 plant species per 1000 individuals, a temperate forest would show approximately 20, and a tropical moist forest would show an astonishing 120 plant species per 1000.[3]

The high diversity of many plant communities presents a practical problem for the use of plant community analysis in the identification and mapping of wetlands. If a community analysis attempts to encompass all species, it may be unreasonably encumbered by the difficulties of dealing with large numbers of rare species. Is it legitimate then to base the diagnosis of wetlands just on the common species?

Common species actually offer some important advantages over rare species for diagnosing and mapping wetlands. A rare species may be represented by very few individuals or even a single organism. The presence of a few individuals could be explained by physical irregularities in the environment. Suppose, for example, that a wetland contained a small mound sufficiently high to be above the zone of prolonged and recurrent saturation. In effect, this would be a small island of upland in a sea of wetland. This island might well support several upland species, each represented by a few individuals. Intensive random sampling would reveal these rare species. To assign much significance to these species would obviously be wrong, given that rarity in this instance would be dictated by the atypical conditions of the microenvironment in which the rare species were found.

Unfortunately for wetland diagnosis, the example of an island of upland in a sea of wetland is not hypothetical. In fact, many wetlands have significant inclusions of upland. Thus, characterization of a community is best when based on its common species, which reflect the predominant environmental conditions.

The greatest difficulty with reliance on common species is that such species by their very nature may be among the most flexible ecologically. Common species of wetlands will inevitably include some species that have a sufficient range of adaptation or a sufficient

3. These are approximations; actual numbers would vary from one location to another (Hubbell 1979).

amount of genetic variation to occupy uplands as well as wetlands (Tiner 1991). In other words, many will be facultative rather than obligate wetland species. The facultative nature of many common wetland species is the key problem of using vegetation analysis for the diagnosis of wetlands.

Present methods of vegetation analysis for the assessment of wetlands are based on the assumption that common species provide sufficient information for classification of ecosystems. The federal government has used two methods for classification: dominance analysis and prevalence indexing. Dominance analysis excludes rare species from consideration. Prevalence indexing includes all species, but, because it weights species according to their abundance, rare species have only small influence on the outcome.

Two pu-dominant multivariate analysis

The Book of Lists

While not authorized to regulate wetlands, the United States Fish and Wildlife Service (USFWS) has been deeply involved in the diagnosis and mapping of wetlands longer than any other federal agency. As explained in chapter 1, the interest of USFWS in wetlands goes back many decades in the context of waterfowl management. The USFWS maintains the only comprehensive maps of wetlands over the entire United States.[4] The mapping project, which led in the 1980s to the first comprehensive estimates of wetland losses, continues in the form of the USFWS National Wetlands Inventory (NWI). The NWI maps, which are produced mainly by aerial photography (Dahl and Johnson 1991), are too general in most cases for use related to the issuance of permits. For this reason, there has been some lack of enthusiasm within the federal government and elsewhere about completion of the mapping project (which has been stalled on a number of occasions due to a lack of funding) and for the continuation of mapping through future cycles that would allow revision of maps sufficient to reflect changes in wetland area (Tiner 1996).

In preparing the National Wetlands Inventory, the USFWS has relied heavily on the assessment of vegetation. Knowing that the species of plants found in wetlands are a subset (about 30%) of all plant

4. The NWI maps were 84% complete for the contiguous 48 states and 29% complete for Alaska as of 1995 (Shafer 1995).

species, the USFWS began to create a list of wetland plants, which came to be called the Hydrophyte List (Reed 1988). The list, which was developed under the direction of Porter Reed of the USFWS, was based upon the experience of botanists regarding the association of particular plant species with wetlands.

The Hydrophyte List proved to be so useful that it developed through several stages after the adoption of wetland regulations in the 1970s. Continuing under the direction of Porter Reed, but with sponsorship from the NRCS and participation of all federal agencies involved in mapping or delineating wetlands, the list was formalized, distributed, and expanded through comprehensive examination of regional floras.

As did the federal soils committee (NTCHS), the federal hydrophyte committee needed to work from a standard definition in preparing its lists of wetland plans. The committee's definition declared that hydrophytes are plants capable of growing in soils that are often or constantly saturated with water during the growing season. Unfortunately, this definition is not fully consistent with the definition of hydric soils: hydrophytes may be found where the substrate is saturated for long intervals during the growing season, but hydric soils can be found only where such saturation has led to the development of low redox conditions in the soil. Thus in principle, one could find federally designated hydrophytes growing on soils that fail the federal test for hydric soils.

The keepers of the Hydrophyte List were aware from the beginning that plants on the list have varying degrees of association with wetlands. In 1982, this information was added to the list (National Research Council 1995). The committee acknowledged 13 regions corresponding to the USDA's regional subdivision of the United States. The committee then asked a selection of botanists in each region to assign a category to each of the species on the Hydrophyte List. Categories include obligate (score of 1, OBL for short), facultative wetland (2, FACW), facultative (3, FAC), and facultative upland (4, FACU). Obligate upland species (5, UPL) were not included in the Hydrophyte List.

In effect, the scoring system assigns extreme scores to the obligate species (1 or 5) and divides the facultative category into three groups. Facultative wetland (FACW) species are found both in wetlands and uplands, but with disproportionate representation in wetlands. The facultative upland (FACU) category shows just the opposite pattern of distribution and the facultative category (FAC) contains species that show approximately equal distribution between wetlands and uplands.

The hydrophyte committee collected information from specialists within each of the 13 regions and searched for agreement among experts within each region. The degree of agreement was quite high, as would be expected if the assignments were made on the basis of scientific inference, which was the underlying supposition. Species were designated for a particular category only if the committee could obtain agreement among the experts. Otherwise, the species was left undesignated.

The collection of separate information for 13 different regions was essential because particular plant species do not always show uniform environmental affinities across regions. In fact, the genetic makeup and physiological characteristics of a given species may vary considerably from region to region.

The Hydrophyte List, as modified by the inclusion of regional lists and fidelity scores (1–5) for each species, has been the engine for botanical identification of wetlands. The means by which the information in the Hydrophyte List can be applied are varied and have been subject to considerable debate, but the underlying foundation for plant community analysis of wetlands is the Hydrophyte List.

Dominance, Prevalence, and Such

Plant ecologists were primed for the assessment of wetland plant communities by decades of debate on the causes of plant community composition. In the early years of plant ecology, one group of botanists (the Clementsians) held that a plant community is a highly deterministic aggregation of species, each of which plays a role that has fixed complementarity to the others, much like the instrument types in a symphony orchestra. The other camp (Gleasonians) embraced an individualistic concept according to which individual species, while possessing particular adaptations, assort with each other more independently, as would the folks gathered around a long traffic light (Golley 1993). Although the debate was never resolved in one fell swoop, the Gleasonians seem to hold the most points according to modern authorities. Meanwhile, the controversy itself has shed much light on the composition of plant communities.

In the process of arguing about the natural rules by which plant species cluster, plant ecologists developed ways of assessing plant communities. The concept of dominance was central to many of these assessment methods. The general idea of dominance in plant communities is well correlated with the common meaning of dominance as applied to any mixture of traits such as colors or cultures. In broad aspect, the dominant species of a plant community are simply those that rank at the top of any quantitative measure of abundance, prominence, or importance.

Judging Predominance through Dominants

Even when armed with the appropriate regional hydrophyte list and a detailed knowledge of the regional flora, a wetland delineator would be at a loss to classify vegetation as either wetland or upland without some sort of decision rule. If wetland plant communities consisted entirely of obligate wetland species, no decision rule would be necessary; but wetland plant communities typically are a mixture of obligate and facultative species. The key is to use information on the abundance of plant species in classifying the plant community as either wetland or upland when the community contains facultative species.

The general basis for a decision rule can be found in the federal definitions or descriptions of wetlands, which reference a "predominance" of hydrophytic vegetation as a characteristic of wetlands (chapter 1). Predominance is not inherently a quantitative concept but has been made quantitative by the application of common sense. The federal convention for judging predominance of hydrophytic vegetation through the examination of dominant species occurs in four steps: (1) list species in decreasing order of abundance; (2) identify the dominants; (3) look up the wetland fidelity class for each of the dominant species; and (4) determine the percentage of species that fall into the following categories: OBL, FACW, and FAC. A decision rule then can be applied: if 50% or more of the dominants are from the categories OBL, FACW, and FAC, then the community is predominantly hydrophytic and is characteristic of wetland. Otherwise, the community is characteristic of upland.

Like excuses, decision rules usually come in mutually supportive clusters. The 50% dominance rule for identification of wetland plant communities must be supported by a rule that divides plants into wetland and non-wetland categories, and also by a rule for deciding which plant species are dominant.[5] The rule for assignment of species to categories calls for lumping of OBL, FACW, and FAC species. This is a gross simplification, given that a plant community having dominants that are all OBL would almost certainly be wetland, whereas one with dominant species that are all FAC would stand only about a 50% chance of being a wetland (because FAC plants are distributed about equally between wetlands and uplands). The rule for deciding which species are dominant is objective in the sense that it is based on a specific cutoff for percent abundance separating dominants from nondominants, but the cutoff itself was identified by professional judgment and is not a reflection of some inherent law of nature. Given that the main decision rule (50% hydrophytes) and its two subsidiary rules (hydrophytes = OBL, FACW, FAC; cutoff rule for dominants) reflect experience and professional judgment rather than cosmic laws, they have been debated at some length.

Even more diverse uncertainties creep into the classification procedure through the collection of field data. The abundance of plants can be measured in several ways[6] and the choice of measurement methods may affect the nature of the conclusions in some cases. Also, enumeration itself is subject to the statistical uncertainties that would apply to any type of data collection.

Critics of wetland regulation often seize on the judgmental nature of decision rules as evidence that procedures for identification and mapping of wetlands are arbitrary. The rules are, however, rational ways of accommodating uncertainty. For the classification of plant communities by analysis of dominants, the degree of certainty is extremely high if all species are OBL or FACW. Where FAC species are important, particularly with some FACU or even a few UPL in the mix of dominants, the degree of certainty falls until it reaches approximately 50%, which is the threshold for the decision rule. Thus, the appropriate use of plant community analysis is conditional on the outcome of the analysis. If the community is strongly dominated by OBL

5. For example, the 50/20 rule equates dominants with all species that contribute to the first 50% of total abundance in a cumulative, ranked list of species in descending order of abundance, plus any species contributing 20% or more of any layer of vegetation.

6. Dominance can be assessed separately for the different layers of vegetation (canopy, shrub, and herb), which then creates a need for decision rules to apply when indications differ by layers.

and FACW, it is almost certainly wetland. If it is dominated by UPL and FACU, it is almost certainly upland. A more mixed group of dominants, with strong elements of FAC, provides a much weaker basis for deduction and calls for the use of other kinds of evidence (chapter 7).

Judging Predominance through Prevalence

The analysis of plant communities through their dominant species seems logical enough, but one cannot escape the feeling that a slightly different approach would be more rigorous. For example, the division of a community into a group of dominant species and a group of nondominant species is a bit crude in that it treats the least abundant dominant as highly significant and the most abundant nondominant, which is only slightly less abundant, as totally insignificant. Also, the analysis of dominants treats FAC species and OBL species as equally significant, even though OBL species show much higher affinity with wetlands than FAC species. Thus, the analysis of dominants seems to discard useful information unnecessarily.

Prevalence indexing is a way of making more complete use of information on plant abundances and their fidelity categories; it sometimes gives results that differ from those of dominance analysis (Wakeley and Lichvar 1997). The collection of data is essentially the same as for analysis of dominants: the analyst tabulates the plant species in a series of plots or transects, calculates the abundance of the species across all transects or all sampling plots, and, using the regional hydrophyte list, assigns a fidelity category to each species. The next step in the analysis differs, however, from analysis of dominants in that there is no attempt to classify individual species as dominant or not dominant. Each species is listed along with its abundances and a number or score corresponding to its fidelity status (OBL = 1, FACW = 2, FAC = 3, FACU = 4, and UPL = 5). The analyst then obtains an abundance-weighted average for the scores across all species. This weighted average score, or prevalence index, will have a value between 1 and 5.

The decision rule for prevalence indexing is straightforward: a prevalence index below 3.0 indicates wetland and an index above 3.0 indicates upland. The index allows the significance of species to vary according to their abundance and takes into account the differences among species of all five fidelity categories (OBL through UPL). The main disadvantage of the index is that it requires more botanical expertise than dominance analysis because rare species contribute to the calculation. For practical purposes, however, errors in the identification of rare species usually are insignificant because of the low abundance of these species and their consequently small effect on the index.

The prevalence index is subject to many of the same uncertainties as the analysis of dominants. Index values that are very low (less than 2.0) provide a high degree of certainty, but index values in the midrange (e.g., 2.5–3.5) indicate the need for caution.

When Plants Forsake Us

Vascular plants show a certain amount of inertia in the face of environmental change. Thus, the characteristics of a plant community at any given instant are reflective of conditions in the recent past but not necessarily those of the immediate present. This places some limits on the interpretation of data from plant communities.

Change from an upland to a wetland hydrologic regime can occur by unnatural means, such as the impoundment of surface waters leading to the flooding of lands that were previously well drained or by natural means, such as flood-induced change in drainage patterns. The establishment of a new hydrologic regime involving prolonged saturation near the surface or outright flooding will eliminate plant species that are intolerant of these conditions. This process is easily observed along highways where the fill for highway construction has blocked or rerouted natural drainage. The highway traveler thus can see highway departments as creators of wetlands, through the evidence of standing dead upland trees that reflect a change in plant community composition following a change in hydrology.[7]

The change from upland to wetland often is obvious because large UPL species die quickly when exposed to wetland hydrology. Even so, a formal community analysis could produce a confusing result because FAC and FACU species would likely persist and would only over a period of years be joined by OBL and FACW species. Thus, the

---

7. Blockage of drainage by highways also can desiccate wetlands downstream, but this is less likely because engineering practice must allow any steady or frequent flow of water to pass under a road.

analyst of such a community would need to know that hydrology had changed recently in order to make any sense of community analysis. In fact, a community analysis in such a situation might be all but useless because of the recency of change in hydrology.

A similar and even more confusing case occurs when a wetland has been dewatered, as can occur by diversion of water upstream or by installation of drainage systems near the wetland. This is a much more important case because of the human tendency to drain wetlands for agriculture or other purposes.

Many wetland species, especially if already established, are quite tolerant of well-drained soils. If they were not, they could not withstand the dry seasons and dry years that inevitably affect wetlands. Thus the drainage of a wetland, which in effect is the creation of an upland, may be accompanied at first by very little change in the plant community. Eventually, however, symptoms of the conversion to upland begin to appear in the form of upland colonizers that compete with the wetland specialists. The validity of a plant community analysis will be extremely low the moment following hydrologic change but will increase with time over a few decades corresponding to the life span of the longest-lived wetland species. Thus once again, the analyst must know at least the recent hydrologic history of the site in order to make proper evaluation of a plant community.

Non-hydrologic disturbances also may confuse the signals from plant communities. Agriculture, which is among the most extensive of biological disturbances, is an extreme example insofar as it typically involves removal or at least suppression of spontaneously growing vegetation. The NRCS acknowledges this issue with a reference to "normal circumstances" in its definition of wetlands: wetlands, according to NRCS, have hydrology and soils that can support and normally do support hydrophytic vegetation. Thus, it is not possible for someone who is inconvenienced by a wetland to eliminate it by use of a chain saw, any more than a barber could eliminate a personality by use of scissors. If vegetation is absent or severely suppressed, a wetland may still exist, but it must be identified by hydrologic conditions or by soils.

A more troublesome case involves various anthropogenic stresses on vegetation that do not necessarily derive from agriculture or even from an intended disturbance of plant communities. Smog, water pollution, and exotic pests are handy examples, but there are many others. Stressors of this type may affect the composition of community and thus may distort a plant community analysis. Fortunately, the re-siliency of plant communities is high, even though selected species may be very sensitive to non-hydrologic changes. Even so, evidence of general stress on a community should be cause for caution in inter-preting the results of a plant community analysis.

Beyond Plants

In the prairie pothole region of South Dakota, a federal scientist took small samples of the soil surface at the margin of a pothole in an area known to be recurrently inundated during spring of most years.[8] He placed the samples in water and obtained in quick order a variety of aquatic invertebrates that hatched from resting eggs of individuals that had grown during a previous season of inundation. In this way, he il-lustrated one means of using biotic indicators other than plant com-munities for the diagnosis and mapping of wetlands.

Certain taxa of algae, mosses, liverworts, aquatic invertebrates, and microbes are associated with inundation or saturation of the substrate. Thus, the possibilities for biotic mapping of wetlands extend well be-yond the use of higher plants. It is unlikely that any group of organ-isms will ever displace vascular plants for the diagnosis of wetlands, but complete fixation on higher plants is unnecessarily restrictive, especially where natural plant communities have been disturbed, are naturally depauperate in species, or provide confusing indications.

Problems with FACS

For either of the two main methods of botanical wetland diagnosis (analysis of dominants and prevalence indexing), the presence of large numbers of species classified as FAC can be quite bothersome, given that FAC species are found as often in upland as in wetland. One way of reducing the influence of these species on wetland diagnosis is to eliminate them entirely from the analysis. An index or procedure that does so is known as a "FAC-neutral" test.

8. Euliss and Mushet (1999). The incubation of samples turned up a wide variety of organisms that hatched from resting eggs or dormant states, including fairy shrimp, oligochaetes, copepods, and water fleas.

The effectiveness of FAC-neutral tests in resolving classification ambiguities has been studied in a few cases. Somewhat counter to intuition, FAC-neutral tests do not appear to be clearly superior to the more standard tests used in diagnosis of wetland communities (e.g., Golet et al. 1993). When small numbers of FAC species are present, exclusion of FAC species obviously has little effect on the conclusions. When large numbers of FAC species are present, hydrologic conditions may lie close to a line that divides wetland from upland. In such a case, no amount of data manipulation can give a strong signal about the classification of the plant community. Those who are confronted with large numbers of FAC species in a community would do best to place greater emphasis on soil or hydrology in the final diagnosis.

There are a few notorious instances in which FAC species dominate environments that are unquestionably wetlands, as demonstrated by their hydrology and soils. Lowlands dominated by red maple in the northeastern United States are a good example (Tiner 1991). FAC species that are commonly found as dominants in wetlands either are excellent generalists or show genetic variety that is subtle or not outwardly evident. This problem needs more study, which may lead to simple ways of identifying wetland genotypes or at least of confirming the breadth of tolerance for some species.

Future Prospects for Tree Counting

The diagnosis of wetlands by use of vegetation analysis is probably reaching its practical limits. There are still some unsolved mysteries involving degrees of regional variation within species that show high amounts of ecological flexibility or genetic variation, but research along these lines seems unlikely to change the fundamental approach that has developed out of traditional methods for community analysis of plants.

Plant community analysis is useful for delineation in most cases and approaches definitive where OBL and FACW species are dominant, but it has certain inherent limitations. Wetlands that contain large percentages of FAC species cannot be securely diagnosed through vegetation. Vegetation analysis is a tool sharp at both ends but blunt in the middle, and will ever be so.

7

EYE OF NEWT

Whoever would identify a wetland objectively must cook up some combination of evidence from hydrology, soil, and vegetation. Many recipes have been proposed, but all have proven unappealing to many and downright nauseating to some. Some favor one ingredient over the other two, while others insist on all three. These differences go beyond mere matters of taste; they relate directly to probability of error and feasibility of practice.

Statistics by a Nicer Name

Use of observations or measurements to make conclusions always involves some probability of error. Statistics is the discipline to which we turn in our attempt to attach probabilities of error to a particular judgment. That judgments about wetlands lie within the reach of statistics seems to be forgotten most of the time.

Statistics, while regarded by many as an ugly discipline, has been reborn in several more comely forms. One of these is *risk analysis*,[1]

1. Also called risk assessment. Risk analysis has been most rigorously applied to factors that degrade human health and to natural hazards. Ecological risk assessment, while still in the formative state, is based on the premise that undesirable ecological conditions, which are also referred to as ecological endpoints, can be induced by human action and that the probability for realization of a given endpoint can be estimated (Bartell et al. 1992).

which involves the assignment of probability to outcomes that society views as undesirable. Given that risk analysis has been recently as much a rage as statistics itself was a generation ago, wetland identification should be steeped in it, but this is not the case.

In arguing over the identification of wetlands, critics of a particular type of evidence are likely to say that it is unreliable. No such generalization is reasonable; the reliability of a particular type of evidence depends on the situation. For example, a plant community strongly dominated by obligate wetland plant species (or equivalently, showing a prevalence index below 2) will support the identification of a wetland with very little risk of error if there has been no recent change in hydrology. The risk thus could be framed as a statement of conditional probability: given that hydrologic conditions have not changed recently, strong dominance of the plant community by obligate wetland species indicates the presence of wetland with a probability exceeding 90%.[2] On the other hand, dominance of facultative species (e.g., prevalence index of 2.5) would be a very different matter. The evidence would indicate wetland but the underlying probability statement might be more as follows: given that hydrologic conditions have not changed recently, the presence of a plant community dominated by facultative wetland species indicates the presence of a wetland with a probability of 50 to 70%. The risk of error for these two determinations is very different, even though in both cases the vegetation indicates presence of wetland.

Principles of risk or probability attach to each of the three major categories of evidence around which wetland determinations must be oriented: vegetation, soils, and hydrology. As we know from watching TV detectives, a combination of mutually supportive modest probabilities can support a high level of certainty. In fact, the combination of evidence from separate sources as a rule of thumb is multiplicative with respect to uncertainty. As a first approximation, a 20% risk of error in evidence on vegetation, when combined with 20% risk of error from a diagnosis of soils, would present a combined probability of error equal to 0.2×0.2, or only 4%, which is much less than the error associated with either type of evidence alone. Thus, risk analysis

2. The percent given here is only hypothetical; probability of error corresponding to specific prevalence values is unknown, but could be developed at least regionally through appropriate kinds of field studies.

shows how weak evidence from one source can be reinforced with weak evidence from another source.[3]

Figure 7-1 illustrates the effect on risk of adding evidence from two categories or three categories to a determination that may be uncertain if based on evidence from only one category. For a single factor, the highest risk of error is 50%, which represents only a random chance of identifying a wetland. This high probability of error is associated with a wetland indicator that registers neutral. An example would be a prevalence index of 3.0 for vegetation analysis. At the other end of the spectrum, a single wetland indicator pegged to the highest possible value presents a probability of error approaching 0% if hydrologic conditions have not changed recently. For a single factor, it is reasonable to assume that there is a smooth, linear continuum representing the relationship between probability of error and strength of indication from a given factor. When a wetland indicator shows only moderate strength, the risk of error may be quite high (e.g., 25% for an indicator showing a value halfway between neutral and maximum strength). The addition of an indicator representing a separate factor reduces error remarkably, provided that the evidence from the second factor is derived independently of the first; addition of evidence from a third factor reduces risk even further (figure 7-1) but could be achieved only by the direct evaluation of hydrology, which typically is not feasible.

Risk can be an argument either for or against the collection of multiple kinds of evidence. Where the diagnosis of vegetation involves substantial risk of error, the obvious remedy is collection of additional evidence from another category. This is an often cited principle, but the reverse principle, which is equally true, is much less seldom embraced: if the evidence from a particular category (e.g., soil or vegetation) supports a conclusion with very low risk of error, it is unnecessary to accumulate other evidence.

Once again, figure 7-1 makes the point. When the strength of a particular wetland indicator approaches its maximum value, the probability of error becomes very low and there is little justification for the addition of a second factor unless the economic stakes are very high.

3. This rule for the combination of probabilities hinges on the assumption that the categories of evidence are assessed independently of each other. Vegetation analysis typically will be independent of the analysis of soil or hydrology. For reasons explained in chapters 4 and 5, however, indications from soils and hydrology may or may not be methodologically independent of each other.

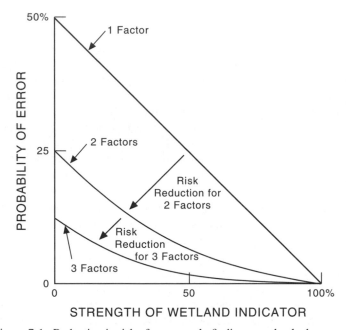

STRENGTH OF WETLAND INDICATOR

Figure 7-1. Reduction in risk of erroneously finding a wetland where none exists, given the use of one, two, or three kinds of evidence. Strength of wetland indicator, as shown on the x-axis, equals zero when the indicator gives a neutral reading (i.e., 50% chance of wetland and 50% chance of upland).

A few students of wetlands have attempted to harness the principle of variable risk in the service of human efficiency. One such attempt is due to Ralph Tiner of the U.S. Fish and Wildlife Service, who has proposed a method that he calls PRIMET (Tiner 1993, 2000). PRIMET involves some very specific methodology, but the underlying principle is that all wetlands have at least one indicator that is diagnostic of the wetland condition. If so, any wetland can be detected simply by the presence of any single diagnostic (primary) indicator, even if this means collecting only a very small amount of evidence. Essentially, this method and others of similar intent make the same assumption that a physician makes upon finding a pulse in an accident victim: the pulse indicates that the patient is alive; further testing is unlikely to contradict this conclusion. The same sort of certainty goes with vegetative communities strongly dominated by obligate wetland plants or underlain by unequivocally hydric soils.

Evidentiary methodology such as that envisioned by Tiner makes consummate sense and is clearly the wave of the future. This wave, however, is not for surfers: it is of diminutive stature because its strength has been sapped by the inertia of established practice.

Cops and Robbers

Underlying our dilatory attitude toward statistical determination of risk in wetland identification is a fundamental disagreement about the significance of risk. The conflict is most easily explained in terms of the judicial analogy of chapter 4. Those who prefer not to find wetlands borrow their approach from criminal law: they abhor uncertainty. Those who want to protect wetlands proceed more on the basis of civil law: they prefer to rely on the weight of evidence, even if the weight of evidence is difficult to judge. In other words, those who prefer to err in excluding wetlands tend to be literal-minded in viewing the evidence, while those who would minimize erroneous exclusions tend to be pragmatic.

The tension between the literalists and pragmatists has not been resolved by law or regulation. The resulting confusion is much the same as if one were attempting to practice law without knowing the standard of proof that might be used in a particular case.

It is hard to have much sympathy with the literalists in the matter of wetlands. Persistent literalism obviously leads to many errors of exclusion for wetlands and promotes the use of technical expertise to contest evidence that is scientifically valid even though not absolutely definitive.

Scientists may sometimes line up with the literalist camp because they are trained to think in terms of alpha error, which is the name applied to errors of commission. For example, a standard test of statistical validity for scientific observations corresponds to a 5% alpha error. In order to make an assertion in print, a scientist is often asked to show an alpha, which is considered good only if very small.

Beta is the error of omission. Scientists are less concerned with beta than alpha, but in this sense, science is not a good role model for resource management. Failure to find something that is actually present (beta error) is commonplace in scientific work, but nobody worries much about it because the individual who is not finding it will be succeeded by many others, one of whom will eventually produce a small

Science is not a good role model for resource management

alpha. This approach will neither work for wetland identification nor for many other matters related to the regulation of resources because the main issue is not to validate a general law or principle, as in basic science, but rather to make a decision on a specific resource that is subject to erroneous misclassification if beta is high. Thus in wetland regulation, beta is at least as important as alpha, and may be even more important if one accepts the argument that loss of wetlands is generally irreversible, whereas erroneous identification of a wetland can always be reversed at a later time by a more comprehensive determination.

Triple Delight

Debate over evidence has been framed not in terms of risk, as it properly should have been, but rather in terms of numbers of factors. The three-factor folks want to see proof related to vegetation, soils, *and* hydrology before admitting that they are looking at a wetland. This group seems to include some sincere technical types who see no reason why the three definitive attributes of wetlands should not all be apparent when a wetland is defined, as well as some more cynical types who view stringent evidentiary requirements as an asset in shackling those who are charged with identifying and delineating wetlands. In fact, the triple-factor requirement is impractical because it requires direct information on hydrology, which is seldom available. The triple-factor approach was the crux of the 1991 proposed revisions, which were never adopted because the gross magnification of beta error in the identification of wetlands was unacceptable to many people (figure 7-2).

Standing just one step below the three-factor folks on the evidentiary staircase are those who insist that they believe in three factors but allow the most difficult factor (hydrology) to be inferred from one or both of the other two factors (vegetation and soils). The distinction between this group and the strict three-factor types is both philosophical and practical. The revisionist three-factor types like the concept of three factors and probably do not want to be identified with zealots who might be suspected of wanting to find as many wetlands as possible, and damn the evidence. The flexible three-factor types are essentially no different from the two-factor types, but the distinction may help them keep better company, as they see it.

The two-factor types are individuals who acknowledge that direct

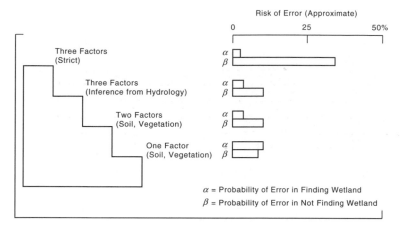

Figure 7-2. Relative risk of alpha and beta error in identifying wetlands as affected by the use of evidence from one, two, or three factors (actual percentages will vary from one site to another).

information on hydrology will seldom be available and that the regulatory system may as well accept the necessity of identifying wetlands mainly on the basis of soils and vegetation. Furthermore, regulatory experience suggests that the resulting compromise in alpha error typically is quite acceptable. Because the hydrologic condition of wetlands is the cause of their hydrophytic vegetation and their hydric soils, it seems a bit superfluous to seek information on hydrology unless hydrology has changed or unless these two other kinds of evidence provide mixed indications.

On the bottom step of the evidentiary ladder is a group of people who view evidence from a single factor as perfectly satisfactory in most instances. Many experts on soils, for example, would feel quite comfortable identifying a wetland on the basis of hydric soil alone, provided that hydrology had not changed. Similarly, strongly predominant hydrophytic vegetation would give the same level of comfort to most botanical experts, but possibly not to soil scientists. The problem with a single-factor philosophy is that its devotees, in order to be practical, must acknowledge the need for evidence from a second or even a third factor whenever the evidence from the first factor is not clear. Thus, rigid reliance on a single factor is no more reasonable than rigid insistence on three factors.

A Search for the Best Recipe

The foregoing consideration of factor factions leads inevitably back to the issue of risk. The proper approach to evidence in the identification of wetlands cannot be framed in terms of numbers of factors because risk of error varies even when the number of factors is constant. Thus, the only rational system is one that specifies the acceptable level of risk and allows the number of factors to vary according to the risk of error in a particular situation. An obvious determination should be possible in some cases from a single factor, but a problematic determination should involve dual factors, and a true borderline case should be decided if at all possible through the use of all three factors.

The prospect of wetland identification and delineation becoming more specifically based on risk is poor because the maturation of regulatory agencies and political factions surrounding the identification of wetlands tends to stabilize the status quo. Unfortunately, technical improvements that can be made easily when a regulatory system is new are very difficult to achieve later, when there is such a thing as established practice. If so, success or failure of the regulatory system will be visible only over intervals of a decade or so. The devil will be in the details.

8

ONCE AND FUTURE WETLANDS

A societal conflict as prolonged and complex as the reversal of national policy on wetlands in the United States must contain some lessons for the future. Perhaps we are still too close to the issues to have everything in perspective historically, but two lessons seem obvious. One of these has to do with the channelizing effect of change in public attitudes toward wetlands and the other with the stabilizing effect of science on regulations and policies intended for the protection of wetlands.

Attitude Adjustment

A look back at the previous chapters suggests that the history of wetland policy in the United States can be divided into three eras: a classical era during which removal was the policy; a modern era during which protection was the policy; and a new era, which appears to be postmodern in the sense that we adjust protection qualitatively in an attempt to make our coexistence with wetlands more comfortable.

Politics of the removal era appear to have been relatively tranquil, as congressional action surrounding wetlands developed almost entirely through consultation with a single interest group (i.e., those who saw some economically beneficial potential in federal progams subsidizing or encouraging the removal of wetlands; Tzoumis 1998). The desire for protection, although present in some circles much earlier, became politically potent in parallel with the growth of general public support for environmental legislation. From that time forward, legislation and national policy have consistently been formed in an atmosphere of strongly opposing viewpoints, but the protectionist impulse has prevailed. It seems doubtful now that an open legislative assault on wetland protection would be successful, simply because the public has fully absorbed the idea of protection for about a generation. The fundamental intent of protectionism, however, still could be subverted judicially or administratively; this is the main issue for the future.

From 1970 to the present, the politics of wetlands has seemed unstable and even chaotic. Participants in the contest over wetlands typically have viewed the future with a high degree of pessimism. This is especially true for the defenders of wetlands, who fear, and in some cases almost anticipate, reactionary backsliding. It is true that protectionism has weathered some credible assaults. A list of these would have to include the 1991 proposed revisions, which would have undermined the protection of wetlands through creation of unreasonable evidentiary requirements. A second was the legislative agenda of the 104th Congress, which was intended to reverse perceived excesses in the protection of wetlands. Both of these failed resoundingly despite the considerable political momentum behind their proponents. Survival of the protectionist agenda appears to be not a reflection of success among its most extreme proponents, but rather of mainstream public sentiment.

Science as Ballast for Policy

The major protagonists in past political wetland wars have been legislators and executives. Far in the background, and often invisible, have been scientists who deal with wetlands. Nowhere in public view has there been an equivalent of the nuclear age's Einstein or Oppenheimer.[1] Even so, a retrospective view shows that scientists have been remarkably influential in facilitating the national change in wetland policy.

The feasibility of protecting wetlands has always depended on ra-

1. Leaders of wetland science, although seldom known to the public, are visible within the science community; most of their names appear in the footnotes of this book.

tional and repeatable means of identifying them. In passing virtually overnight from a situation in which definitions of wetlands were of no general interest whatsoever to one in which large segments of society might be drastically affected by the application of such a definition, the government drew upon a trust fund of scientific expertise about wetlands that had been developing for decades. The sources were many, but perhaps the best example is work leading to the publication by Shaw and Fredine (1956) on wetland types (chapter 2). This work, which was already about 20 years old by the time it found its most widespread use, and similar work that succeeded it, spared the nation of much error and inconsistency that could have resulted from hasty and naive assumptions about the breadth of wetland types and the diagnostic characteristics of wetlands.

Additional scientific preparedness was evident as the Army Corps turned to federal scientists, including especially those at the USACE Waterways Experiment Station in Vicksburg, Mississippi, for principles to be used in mapping wetlands. Scientific consensus developed rapidly around the three-factor approach mentioned in chapter 2. The broad uniformity of acceptance for this approach among the scientific community has been one of the most important stabilizers of wetland regulation in the United States; the speed of its consolidation can be explained only by the mature expertise of a small cadre of individuals who already knew a great deal about wetlands when wetlands became politically important.

Other landmarks of scientific contribution were the formation and subsequent work of the NTCHS (chapter 5) and the creation of the Hydrophyte List (chapter 6). In both cases, scientific knowledge that had been developing over decades in fields often considered obscure or specialized was mobilized for new purposes in a remarkably short time. In each instance, the result was creation of rationale and objectivity around procedures that were necessary for implementation of a controversial new policy. Without the ballast provided by such a weighty synthesis of information from individuals immersed primarily in information rather than policy or politics, the creation of defensible procedures for identifying and delineating wetlands would not have been possible.

An era in the history of wetlands policy debate closed in 1995 with publication of the NRC's wetlands report (chapter 1). In response to charges that wetland regulation was poorly founded and often capricious, Congress called for an independent evaluation and the result was formation of the NRC Committee on Characterization of Wetlands. The committee, while diversely composed with respect to viewpoint and expertise, concluded that the government's basis for regulating wetlands had been fundamentally rational, rested on a sound scientific foundation, and had been implemented in a reasonable way. The committee also found weaknesses and loopholes, some of which it believed to be quite significant, but the ultimate prescription of the committee was for improvement and reform (some of which has since occurred) rather than elimination or structural overhaul of the system. The main reason for the committee's conclusion was the scientific credibility of procedures that had been developed for identifying and mapping wetlands.

Major legislative assaults on wetland protection ceased following the 104th Congress and the publication of the NRC's wetlands report. Such attacks may return, but the main struggle relevant to the future of wetlands in the United States seems to have shifted to entirely different ground.

Litigation and Mitigation

The legal basis for administration of Section 404 has never seemed very secure, given the many steps that separate legislation from permitting and enforcement. Takings issues are the constant focus of attention for many legal experts; these issues, if found in some sustained way by courts to have merit, could undo the present regulatory system (chapter 1). Less fatal, but still very damaging, would be loss of the Tulloch Rule, which is the main means by which drainage of wetlands is prevented (chapter 1). It would be hard to predict in such a case if protectionism could muster enough political support to create legislation that would offset losses through litigation.

Although litigation may be the most potent threat to protection of wetlands, mitigation also presents a current but subtle problem. As explained in chapter 3, mitigation, as applied to wetlands, is a very broad concept in that it could include any component of a plan for environmental alteration that is intended to reduce the loss of area or functional capacity of wetlands. Mitigation has been part of the agenda for Section 404 permitting since the beginning of wetland regulation. In fact, the Section 404 program has succeeded in protecting wetlands not mainly through the denial of permits, but rather through the preference of potential permit seekers to stay away from wetlands in order to avoid delay caused by permitting and to escape potentially costly mitigation that might be extracted from them through permitting.

In its simplest form, mitigation involves negotiations over the design of projects and leads to changes in plan that reduce intrusion on wetlands, establish buffer zones, set aside certain wetlands in perpetuity, etc. In addition, however, mitigation has extended to the restoration of damaged wetlands or former wetlands in exchange for a permit allowing the elimination of undamaged wetlands and, as the final logical extension, substitution of an entirely new wetland (*created wetland*) for a wetland that is damaged or destroyed under permit.

Mitigation that allows substitution of wetland to be destroyed or altered with wetland to be restored or created de novo is increasingly popular (Sapp 1994). On the regulatory side, one-to-one replacement of lost wetland with restored wetland or new wetland seems consistent with a national policy of *no net loss*. In fact, replacement typically is required at a ratio higher than one-to-one, thus seeming to serve the even more ambitious goal of increasing the amount of wetlands in the United States. On the side of development interests, mitigation that involves replacement has become attractive through the evolution of a system known as mitigation banking. The mitigation banker puts up financial resources to restore or, more typically, to create wetland from scratch and puts the new wetland acreage up for sale as mitigation for destruction of other wetlands under permit. In order to succeed, the wetland banker must establish credibility with the regulatory agency (which may require monitoring of wetland status), assurances of continuity, or other measures designed to maximize the validity of the exchange.[2]

Mitigation through the creation of new wetlands developed slowly at first because this type of mitigation is mainly achieved through the private sector, which needed time to test the concept. The concept has proven sound from several viewpoints. For example, the idea of a shoe company or a tire factory trying to restore or create wetlands on its own, even with the help of consultants, is troublesome, but mitigation banking makes it seem feasible. Consolidation of mitigation through mitigation bankers allows regulatory agencies to make more extensive

---

2. The bank in most cases is private, but public agencies may serve as bankers through *in lieu fee* programs (Stein 1999). Banks function under interagency regulations adopted in 1995 (Federal Register 60: 58605 [1995]).

requirements, expect a higher level of technical expertise, and deal with larger wetlands than would be possible through negotiations with most individual permit holders.

Given the advantages of mitigation banking or other mechanisms for creating wetlands de novo as a means of mitigation, it seems perverse to raise objections. Judging from the history of wetlands regulation, however, extensive use of such a scheme is ill-advised unless it rests on a firm scientific foundation. There is much doubt as to whether this foundation exists.[3]

As explained in chapter 3, there is more than a speck of uncertainty in the notion that wetlands can be replaced functionally, even if the hydrologic conditions for existence of a wetland are provided with certainty (Zedler 1996). The main problem is that many of the functions of wetlands are conducted by or with the assistance of organisms, and organisms are exquisitely sensitive to environmental conditions; they simply will not grow unless the complete matrix of conditions to which they are adapted is available. Furthermore, we are woefully ignorant of the essential conditions even for the most common of wetland organisms and therefore face great difficulty in taking a methodical approach to satisfying the requirements of even one kind of organism. We are left with a default approach in which we provide certain hydrologic conditions (but often with too little attention to the natural regime [Bedford 1996]), do some planting of hydrophytes that we would like to see established, and hope everything turns out for the best. Certainly, numerous wetland organisms will appear, but they may not include the same kinds of organisms that were present in the wetland that was eliminated and they may not be able to perform the same suite of functions that were eliminated with the original wetland (Middleton 1999). Thus, this type of mitigation poses a risk that is presently very difficult to evaluate.

The early scientific challenges of wetland regulation involved identification and mapping of wetlands and analyzing wetland functions. The scientific community was prepared for these challenges not by design, but by accident through the fortunate practice we have of maintaining at least a few individuals in virtually every field of science. The scientific expertise that bolstered wetland regulations from the 1970s

---

3. The National Research Council has formed a committee to study the scientific basis for mitigation. The committee's report will be an important guide as to the current feasibility of mitigation.

into the 1990s for the most part was simply mobilized and rearranged; it was not created from scratch. The present situation with regard to mitigation is very different in that the study of restoration and replacement is relatively new and undeveloped (National Research Council 1992).

The great danger in mitigation through restoration or establishment of artificial wetlands lies in the ease with which these practices could serve as a universal escape valve for pressures promoting the protection of natural, undamaged wetlands. When restored wetlands or new wetlands are weighed against no mitigation whatsoever, they seem desirable, but not so when they are weighed against unperturbed natural wetlands. Until someone shows how environmental engineering can provide us with wetlands that are indistinguishable from the natural ones, there is a penalty associated with mitigation through replacement. In fact, mitigation through replacement will probably always be to varying extents inadequate, in which case the most reliable means to *no net loss* may be just as we originally thought: protection of wetlands as we find them.

REFERENCES

Altseri, M. A. 1995. *Agroecology: The Science of Sustainable Agriculture.* 2nd ed. Westview Press, Boulder, Colo.

Anfinson, J. O. 1995. Floodplain ecosystem change in the Mississippi River. Pp. 369–377 in J. A. Kusler, D. E. Willard, and H. C. Hull Jr. (eds.), *Wetlands and Watershed Management.* Association of State Wetland Managers, Berne, N.Y.

Angermeier, P. L., and J. R. Karr. 1994. Biological integrity versus biological diversity as policy directives. *Bioscience* 44:690–697.

Babcock, H. 1991. Federal wetlands regulatory policy: Up to its ears in alligators. *Pace Environmental Law Review* 6:307–353.

Baker, H. G. 1989. Sources of the naturalized grasses and herbs in California grasslands. Pp. 29–38 *in* L. F. Huenneke and H. Mooney (eds.), *Grassland Structure and Function.* Kluwer, Dordrecht, Netherlands.

Bartell, S. N., R. H. Gardner, and R. V. O'Neill. 1992. *Ecological Risk Estimation.* Lewis Publishers, Boca Raton, Fla.

Bedford, B. L. 1996. The need to define hydrologic equivalents at the landscape scale for freshwater wetland mitigation. *Ecological Applications* 6:57–68.

Bedford, B. L., M. B. Brinson, R. Sharitz, A. Vander Valk, and J. Zedler. 1992. Evaluation of proposed revisions to the 1989 "Federal Manual for Identifying and Delineating Jurisdictional Wetlands." *Bulletin of the Ecological Society of America* 73:14–23.

Belden and Russonello, Inc. 1996. Human values and nature's future: Americans' attitudes on biological diversity. United States Fish and Wildlife Service document (unpublished).

135

Birkeland, P. W. 1984. *Soils and Geomorphology*. Oxford University Press, New York, N.Y.

Birren, F. 1969. *A Grammar of Color*. Van Nostrand Reinhold, New York, N.Y.

Bocking, S. 1997. *Ecologists and Environmental Politics: A History of Contemporary Ecology*. Yale University Press, New Haven, Conn.

Bosselman, F. P. 1996. Limitations inherent in the title to wetlands at common law. *Stanford Environmental Law Journal* 15:247–337.

Brady, N. C., and R. R. Weil. 1996. *The Nature and Properties of Soils*. 11th ed. Prentice-Hall, Upper Saddle River, N.J.

Briggs, A. 1983. *A Social History of England*. Viking Press, New York, N.Y.

Brinson, M. M. 1993. *A Hydrogeomorphic Classification for Wetlands*. Wetlands Research Program Technical Report WRP-DE-4. U.S. Army Corps of Engineers, Waterways Experiment Station. Vicksburg, Miss.

———. 1995. *Guidebook for Application of Hydrogeomorphic Assessments to Riverine Wetlands*. Report WRP-DE-11. U.S. Army Corps of Engineers, Waterways Experiment Station. Vicksburg, Miss.

———. 1998. More clarification regarding the HGM approach. *Bulletin of the Society of Wetland Scientists* 15:7–10.

Burdick, D. E. M., D. Cushman, R. Hamilton, and J. G. Gosselink. 1989. Faunal changes in bottomland hardwood forest loss in the Tensas watershed, Louisiana. *Conservation Biology* 3: 282–292.

Canadell, J., R. B. Jackson, J. R. Ehleringer, H. A. Mooney, O. E. Sala, and E. D. Schulze. 1996. Maximum rooting depth of vegetation types at the global scale. *Oecologia* 108:583–595.

Christensen, N. L., A. M. Bartuska, N. J. Brown, S. Carpenter, C. D'Antonio, R. Francis, J. F. Franklin, J. A. MacMahon, R. F. Noss, D. J. Parsons, C. H. Peterson, M. G. Turner, and R. G. Woodmansee. 1996. Report of the Ecological Society of America Committee on Scientific Basis for Ecosystem Management. *Ecological Applications* 65:665–691.

Clark, E. H., II. 1993. Three little words: Understanding and implementing a "no-net-loss" goal. Prepared for the First National Wildlife Habitat Workshop, Winnipeg, Manitoba, Canada.

Cohen, J. E., and D. Tilman. 1996. Biosphere 2 and biodiversity: The lesson so far. *Science* 1150–1151.

Conservation Foundation. 1988. *Protecting America's Wetlands: An Action Agenda*. Final report of the National Wetlands Policy Forum. Washington, D.C. Conservation Foundation, Washington, D.C.

Cowardin, L. M., V. Carter, F. C. Golet, and E. L. Lareau. 1979. *Classification of wetlands and deep water habitats of the United States*. FWS/OBS 79/31. U.S. Fish and Wildlife Service, Washington, D.C.

Dahl, T. E. 1990. *Wetlands Losses in the United States, 1780s to 1980s*. U.S. Department of the Interior, Fish and Wildlife Service, Washington, D.C.

Dahl, T. E., and C. E. Johnson. 1991. *Wetland Status and Trends in the Coter-*

minous United States, Mid-1970's to Mid-1980's. First update of the National Wetlands Status Report. U.S. Fish and Wildlife Service, Washington, D.C.

Dahl, T. E. 2000. *Status and Trends of Wetlands in the Conterminous United States, 1986 to 1997.* U.S. Department of the Interior, Fish and Wildlife Service, Washington, D.C.

DeLaune, R. D., R. R. Boar, C. W. Liwdau, and B. A. Kleiss. 1996. Denitrification in bottomland hardwood wetlands of the Cache River. *Wetlands* 16:309–320.

Dooge, J. C. I. 1984. The waters of the earth. *Hydrologic Science Journal* 29:149–176.

Ducks Unlimited. Summer, 1997. *American Waterfowl Timeline.* p.117.

EPA. 1994. *Corps of Engineers, Natural Resource Conservation Service, Fish and Wildlife Service Memorandum of Agreement Concerning Wetland Determinations on Agricultural Lands.* Washington, D.C.

Euliss, N. H., Jr., and D. M. Mushet. 1999. Influence of agriculture on communities of temporary wetlands in the prairie pothole regions of North Dakota, U.S.A. *Wetlands* 19:578–583.

Ewel, K. C. 1990. Swamps. Pp. 281–322 in R. L. Myers and J. J. Ewel (eds.), *Ecosystems in Florida.* University of Central Florida Press, Orlando, Fla.

Fenchel, T., and B. J. Finlay. 1995. *Ecology and Evolution of Anoxic Worlds.* Oxford University Press, New York, N.Y.

Finder, J., and S. M. Reiness. 1997. General permits under wetlands law. The rise and fall of Nationwide Permit 26. *Environmental Lawyer* 3:891–910.

Fink, R. J. 1994. The national wildlife refuges: Present trends and prospects. *Harvard Environmental Law Review* 18:1–135.

Forbes, S. A. 1887. The Lake as a Microcosm. *Peoria (Illinois) Science Association Journal,* 77–87.

Gates, D. M. 1993. *Climate Change and its Biological Consequences.* Sinauer Associates, Sunderland, Mass.

Golet, F. C., A. J. K. Calhoun, W. R. DeRagon, D. J. Lowry, and A. J. Gold. 1993. *Ecology of Red Maple Swamps in the Glaciated Northeast: A Community Profile.* Biological Report 12. U.S. Fish and Wildlife Service, Washington, D.C.

Golley, F. B. 1993. *A History of the Ecosystem Concept in Ecology.* Yale University Press, New Haven, Conn.

Gosselink, J. G., G. P. Shaffer, L. C. Lee, D. M. Burdick, D. L. Childers, N. C. Leibowitz, S. C. Hamilton, R. Boumans, D. Cushman, S. Fields, M. Koch, and J. M. Visser. 1990. Landscape conservation in a forested wetland watershed. *Bioscience* 40:588–600.

Helfield, J. M., and M. L. Diamond. 1997. Use of constructed wetlands for urban stream restoration: A critical analysis. *Environmental Management* 21:329–341.

Hopkins, B. 1957. The concept of minimal area. *Journal of Ecology* 45: 441–449.

Hornberger, G. M., J. P. Raffensperger, P. L. Wiberg, and K. N. Eshleman. 1998. *Elements of Physical Hydrology*. Johns Hopkins University Press, Baltimore, Md.

Hubbell, S. P. 1979. Tree dispersion, abundance, and diversity in a tropical dry forest. *Science* 203:1299–1309.

Hurt, G. W., and W. E. Puckett. 1992. Proposed hydric soil criteria and their field identification. Pp. 148–151 in J. M. Kimble (ed.), *Proceedings of the 8th International Soil Correlation Meeting: Characterization, Classification, and Utilization of Wet Soils*. USDA, Soil Conservation Service, National Soil Survey Center, Lincoln, Nebr.

Iltis, A. 1975. Phytoplancton des eaux natronées du Kanem (Tchad). Conclusion (1). Office de la Recherche Scientifique et Technique Outre-Mer, Cahiers Série Hydrobiologie. 9:13–17.

Johnston, C. A. 1991. Sediment and nutrient retention by freshwater wetlands: Effects on surface water quality. *CRC Critical Reviews in Environmental Control* 21:491–565.

Leith, H. 1973. Primary production. *Human Ecology* 1:303–332.

Lewis, W. M., Jr. 1994. The ecological sciences and the public domain. *University of Colorado Law Review* 65:279–292.

Martin, J. H., W. H. Leonard, and D. L. Stamp. 1976. *Principles of Field Crop Production*. 3rd ed. Macmillan, New York, N.Y.

McElfish, J. N. 1994. Property rights, property roots: Rediscovering the basis for legal protection of the environment. *Environmental Law Reporter* 24:10,231–10,249.

McIntosh, R. P. 1985. *The Background of Ecology: Concept and Theory*. Cambridge University Press, New York, N.Y.

Middleton, B. 1999. *Wetland Restoration, Flood Pulsing, and Disturbance Dynamics*. John Wiley and Sons, New York, N.Y.

Mitsch, W. J., and J. G. Gosselink. 2000. *Wetlands*. 3rd ed. John Wiley and Sons, New York, N.Y.

Myers, N. 1983. *A Wealth of Wild Species: A Storehouse for Human Welfare*. Westview Press, Boulder, Colo.

National Food Security Act Manual. 1994. Part 519, 180-V-NFSAM. 3rd ed. March 1994. Soil Conservation Service, USDA, Washington, D.C.

National Research Council. 1992. *Restoration of Aquatic Ecosystems*. National Academy Press, Washington, D.C.

———. 1995. *Wetlands: Characteristics and Boundaries*. National Academy Press, Washington, D.C.

Patrick, R., F. Douglass, D. M. Palavage, and P. M. Stewart. 1992. *Surface Water Quality: Have the Laws Been Successful?* Princeton University Press, Princeton, N.J. 198 pp.

Phillips, J. C., and F. C. Lincoln. 1930. *American Waterfowl. Their Present Situation and the Outlook for the Future.* Houghton Mifflin, Boston, Mass.

Pimentel, D. 1984. Energy flow in the food system. Pp. 1–24 in D. Pimentel and C. W. Hall (eds.), *Food and Energy Sources.* Academic Press, New York, N.Y.

Prosser, C. L. (ed.). 1991. *Environmental and Metabolic Animal Physiology.* Wiley-Liss, New York, N.Y.

Raven, P. H. 1988. Our diminishing tropical forests. Pp. 119–122 in E. O. Wilson (ed.), *Biodiversity.* National Academy Press, Washington, D.C.

Reddy, K. R., and E. M. D'Angelo. 1994. Soil processes regulating water quality in wetlands. Pp. 309–324 in W. J. Mitsch (ed.), *Global Wetlands: Old World and New.* Elsevier Science, New York, N.Y.

Reed, P. B. 1988. *National List of Plant Species that Occur in Wetlands: National Summary.* Biological Report 88(24). U.S. Fish and Wildlife Service, Washington, D.C.

Rosenzweig, M. L. 1995. *Species Diversity in Space and Time.* Cambridge University Press, New York, N.Y.

Sapp, W. W. 1994. Mitigation banking. *Environmental Lawyer* 1:99–139.

Schlegel, H. G. 1993. *General Microbiology.* 2nd ed. Cambridge University Press, New York, N.Y.

Schlesinger, W. H. 1997. *Biogeochemistry: An Analysis of Global Change.* 2nd ed. Academic Press, New York, N.Y.

Shafer, L. J. 1995. An overview and status report on NWI digital wetlands data. Pp. 101–102 in J. A. Kusler, D. E. Willard, and H. C. Hull Jr. (eds.), *Wetlands and Watershed Management.* Association of State Wetland Managers, Berne, N.Y.

Shallat, T. 1994. *Structures in the Stream. Water, Science, and the Rise of the U.S. Army Corps of Engineers.* University of Texas Press, Austin, Tex.

Shaw, S. T., and C. G. Fredine. 1956. *Wetlands of the United States: Their Extent, and Their Values for Waterfowl and Other Wildlife.* Circular 39. U.S. Fish and Wildlife Service, Washington, D.C.

Siegel, D. I. 1988. The recharge–discharge functions of wetlands near Juneau, Alaska. Parts I and II. *Groundwater* 26:427–434 and 580–586.

Simonson, R. W. 1997. Evolution of soil series and type concepts in the United States. Pp. 79–108 in D. H. Yaalon and S. Berkowicz (eds.), *History of Soil Science.* Catena Verlag, Reiskirchen, Germany.

Slocombe, S. 1993. Implementing ecosystem-based management. *Bioscience* 43:612–622.

Snyder, D. 1995. What farmers should know about wetlands. *Journal of Soil and Water Conservation* 50:630–632.

Soil Survey Staff. 1975. *Soil Taxonomy.* USDA Soil Conservation Service Agricultural Handbook #436. U.S. Government Printing Office, Washington, D.C.

————. 1992. *Keys to Soil Taxonomy*. 6th ed. USDA, Washington, D.C.

Solley, W. B., and R. R. Pierce. 1992. Preliminary estimates of water use in the United States, 1990. *U.S. Geological Survey Open File Report* 92-63.

Stein, E. D. 1999. Mitigation Banking: Challenges and Lessons Learned. *Bulletin of the Society of Wetlands Scientists* 16:18–22.

Strand, M. 1993. Federal Wetlands Law. *Environmental Law Reporter* 23:10,185–10,215 and 10,354–10,378.

Tansley, A. G. 1935. The use and abuse of vegetational concepts and terms. *Ecology* 16:284–307.

Tilman, D. 1999. The ecological consequences of changes in biodiversity: A search for general principles. *Ecology* 80:1455–1474.

Tiner, R. 1991. The concept of hydrophyte for wetland identification. *Bioscience* 41:236–247.

————. 1993. The primary indicators method—A practical approach to wetland recognition and delineation in the United States. *Wetlands* 13:50–64.

————. 1996. NWI maps—Basic information on the nation's wetlands. *Bioscience* 47:269–271.

————. 1998. *In Search of Swampland*. Rutgers University Press, New Brunswick, N.J.

————. 2000. A review of wetland identification and delineation techniques, with recommendations for improvements. *Wetland Journal* 12:15–22.

Tzoumis, K. A. 1998. Wetland policymaking in the U.S. Congress from 1789 to 1995. *Wetlands* 18:447–459.

U.S. Army Corps of Engineers. 1987. *USACE Wetlands Delineation Manual*. Environmental Laboratory, U.S. Army Engineers Waterways Experiment Station Technical Report Y-87-1. Vicksburg, Miss.

USDA Natural Resource Conservation Service. 1996. *A Geography of Hope*. Publication 1548. Washington, D.C.

USDA Soil Conservation Service. 1980. *Soil Survey of Weld County, CO. Southern Part*. Washington, D.C.

USEPA. 1990. *Environmental Investments: The Cost of a Clean Environment*. Island Press, Washington, D.C.

Valentyne, J. R. 1974. *The Algal Bowl—Lakes and Man*. Ottawa Miscellaneous Publication 22. Ottawa Department of the Environment, Ontario, Canada. 185 pp.

Wakeley, J. S., and R. W. Lichvar. 1997. Disagreements between plot-based prevalence indices and dominance rank in evaluations of wetland vegetation. *Wetlands* 17:301–309.

Wilen, B. O., and R. W. Tiner. 1993. Wetlands of the United States. Pp. 515–636 in D. F. Whigham, D. Dagmar, and H. Slavomil (eds.), *Wetlands of the World*. I. Kluwer, Amsterdam, Netherlands.

Wilson, E. O. 1984. *Biophilia*. Harvard University Press, Cambridge, Mass.

————. (ed.). 1988. *Biodiversity*. National Academy Press, Washington, D.C.

Wilson, M. A., and S. R. Carpenter. 1999. Economic valuation of freshwater ecosystem services in the United States: 1971–1997. *Ecological Applications* 9:772–783.

Winter, T. C., J. W. Harvey, O. L. Franks, and W. M. Alley. 1998. *Ground Water and Surface Water: A Single Resource.* U.S. Geological Survey Circular 1139. Denver, Col.

Wurtsbaugh, W. A. 1992. Food-web modification by an invertebrate predator in the Great Salt Lake. *Oecologia* 89:168–175.

Zedler, J. B. 1996. Ecological issues and wetland mitigation: An introduction to the forum. *Ecological Applications* 6:33–37.

INDEX